Problems and Solutions
in Medical Physics

Series in Medical Physics and Biomedical Engineering

Series Editors: John G. Webster, E. Russell Ritenour, Slavik Tabakov, and Kwan Hoong Ng

Recent books in the series:

Clinical Radiotherapy Physics with MATLAB®: A Problem-Solving Approach
Pavel Dvorak

Advances in Particle Therapy: A Multidisciplinary Approach
Manjit Dosanjh and Jacques Bernier (Eds)

Radiotherapy and Clinical Radiobiology of Head and Neck Cancer
Loredana G. Marcu, Iuliana Toma-Dasu, Alexandru Dasu, and Claes Mercke

Problems and Solutions in Medical Physics: Diagnostic Imaging Physics
Kwan Hoong Ng, Jeannie Hsiu Ding Wong, and Geoffrey D. Clarke

Advanced and Emerging Technologies in Radiation Oncology Physics
Siyong Kim and John W. Wong (Eds)

A Guide to Outcome Modeling in Radiotherapy and Oncology: Listening to the Data
Issam El Naqa (Ed)

Advanced MR Neuroimaging: From Theory to Clinical Practice
Ioannis Tsougos

Quantitative MRI of the Brain: Principles of Physical Measurement, Second Edition
Mara Cercignani, Nicholas G. Dowell, and Paul S. Tofts (Eds)

A Brief Survey of Quantitative EEG
Kaushik Majumdar

Handbook of X-ray Imaging: Physics and Technology
Paolo Russo (Ed)

Graphics Processing Unit-Based High-Performance Computing in Radiation Therapy
Xun Jia and Steve B. Jiang (Eds)

Targeted Muscle Reinnervation: A Neural Interface for Artificial Limbs
Todd A. Kuiken, Aimee E. Schultz Feuser, and Ann K. Barlow (Eds)

Emerging Technologies in Brachytherapy
William Y. Song, Kari Tanderup, and Bradley Pieters (Eds)

Environmental Radioactivity and Emergency Preparedness
Mats Isaksson and Christopher L. Rääf

The Practice of Internal Dosimetry in Nuclear Medicine
Michael G. Stabin

Problems and Solutions in Medical Physics
Nuclear Medicine Physics

Kwan Hoong Ng

Chai Hong Yeong

Alan Christopher Perkins

CRC Press
Taylor & Francis Group
Boca Raton London New York

CRC Press is an imprint of the
Taylor & Francis Group, an **Informa** business

CRC Press
Taylor & Francis Group
6000 Broken Sound Parkway NW, Suite 300
Boca Raton, FL 33487-2742

© 2019 by Taylor & Francis Group, LLC
CRC Press is an imprint of Taylor & Francis Group, an Informa business

No claim to original U.S. Government works

Printed on acid-free paper

International Standard Book Number-13: 978-1-4822-4000-9 (Paperback)
978-0-367-14797-6 (Hardback)

Visit the Taylor & Francis Web site at
http://www.taylorandfrancis.com

and the CRC Press Web site at
http://www.crcpress.com

Contents

About the series

The *Series in Medical Physics and Biomedical Engineering* describes the applications of physical sciences, engineering and mathematics in medicine and clinical research.

The series seeks (but is not restricted to) publications in the following topics:

- Artificial organs
- Assistive technology
- Bioinformatics
- Bioinstrumentation
- Biomaterials
- Biomechanics
- Biomedical engineering
- Clinical engineering
- Imaging
- Implants
- Medical computing and mathematics
- Medical/surgical devices
- Patient monitoring
- Physiological measurement
- Prosthetics
- Radiation protection, health physics and dosimetry
- Regulatory issues
- Rehabilitation engineering
- Sports medicine
- Systems physiology
- Telemedicine
- Tissue engineering
- Treatment

THE INTERNATIONAL ORGANIZATION FOR MEDICAL PHYSICS

The International Organization for Medical Physics (IOMP) represents over 18,000 medical physicists worldwide and has a membership of 80 national and 6 regional organisations, together with a number of corporate members. Individual medical physicists of all national member organisations are also automatically members.

The mission of IOMP is to advance medical physics practice worldwide by disseminating scientific and technical information, fostering the educational and professional development of medical physics and promoting the highest quality medical physics services for patients.

A World Congress on Medical Physics and Biomedical Engineering is held every three years in cooperation with International Federation for Medical and Biological Engineering (IFMBE) and International Union for Physics and Engineering Sciences in Medicine (IUPESM). A regionally based international conference, the International Congress of Medical Physics (ICMP) is held between world congresses. IOMP also sponsors international conferences, workshops and courses.

The IOMP has several programmes to assist medical physicists in developing countries. The joint IOMP Library Programme supports 75 active libraries in 43 developing countries, and the Used Equipment Programme coordinates equipment donations. The Travel Assistance Programme provides a limited number of grants to enable physicists to attend the world congresses.

IOMP co-sponsors the *Journal of Applied Clinical Medical Physics.* The IOMP publishes, twice a year, an electronic bulletin, *Medical Physics World.* IOMP also publishes e-Zine, an electronic newsletter, about six times a year. IOMP has an agreement with Taylor & Francis Group for the publication of the *Medical Physics and Biomedical Engineering* series of textbooks. IOMP members receive a discount.

IOMP collaborates with international organisations, such as the World Health Organization (WHO), the International Atomic Energy Agency (IAEA) and other international professional bodies, such as the International Radiation Protection Association (IRPA) and the International Commission on Radiological Protection (ICRP), to promote the development of medical physics and the safe use of radiation and medical devices.

Guidance on education, training and professional development of medical physicists is issued by IOMP, which is collaborating with other professional organisations in development of a professional certification system for medical physicists that can be implemented on a global basis.

The IOMP website (www.iomp.org) contains information on all the activities of the IOMP, policy statements 1 and 2, and the 'IOMP: Review and Way Forward', which outlines all the activities of IOMP and plans for the future.

Preface

In view of the increasing number and popularity of master's and higher-level training programmes in medical physics worldwide, there is an increasing need for students to develop problem-solving skills in order to grasp the complex concepts which are part of ongoing clinical and scientific practice. The purpose of this book is, therefore, to provide students with the opportunity to learn and develop these skills.

This book serves as a study guide and revision tool for postgraduate students sitting for examinations in nuclear medicine physics. The detailed problems and solutions included in the book cover a wide spectrum of topics, following the typical syllabi used by universities on these courses worldwide.

The problems serve to illustrate and augment the underlying theory and provide a reinforcement of basic principles to enhance learning and information retention. No book can claim to cover all topics exhaustively, but additional problems and solutions will be made available periodically on the publisher's website: https://crcpress.com/9781482239959.

One hundred and forty-eight solved problems are provided in ten chapters on nuclear medicine physics topics, including radioactivity and nuclear transformation, radionuclide production and radiopharmaceuticals, non-imaging detectors and counters, instrumentation for gamma imaging, SPECT and PET/CT imaging techniques, radionuclide therapy, internal radiation dosimetry, quality control and radiation protection in nuclear medicine. The approach to the problems and solutions covers all six levels in the cognitive domain of Bloom's taxonomy.

This book is one of a three-volume set containing medical physics problems and solutions. The other two books in the set tackle diagnostic imaging physics and radiotherapy physics.

We would like to thank the staff at Taylor & Francis Group, especially Francesca McGowan and Rebecca Davies, for their unfailing support.

Kwan Hoong Ng, Chai Hong Yeong, Alan Christopher Perkins

Authors

Kwan Hoong Ng, PhD, FinstP, DABMP, received his MSc (medical physics) from the University of Aberdeen and PhD (medical physics) from the University of Malaya, Malaysia. He is certified by the American Board of Medical Physicists. Professor Ng was honoured as one of the top 50 medical physicists in the world by the International Organization of Medical Physics (IOMP) in 2013. He also received the International Day of Medical Physics Award in 2016. He has authored/co-authored over 230 papers in peer-reviewed journals and 25 book chapters and has co-edited 5 books. He has presented over 500 scientific papers and has more than 300 invited lectures. He has also organised and directed several workshops on radiology quality assurance, digital imaging and scientific writing. He has directed research initiatives in breast imaging, intervention radiology, radiological safety and radiation dosimetry. Professor Ng serves as a consultant for the International Atomic Energy Agency (IAEA) and is a member of the International Advisory Committee of the World Health Organization (WHO), in addition to previously serving as a consulting expert for the International Commission on Non-Ionizing Radiation Protection (ICNIRP). He is the founding and emeritus president of the South East Asian Federation of Organizations for Medical Physics (SEAFOMP) and is a past president of the Asia-Oceania Federation of Organizations for Medical Physics (AFOMP).

Chai Hong Yeong, PhD, is a medical physicist and an associate professor at the School of Medicine, Taylor's University, Subang Jaya, Malaysia. Dr. Yeong received her BSc degree in health physics in 2005, master of medical physics in 2007, and PhD in medical physics in 2012. She is currently a council member of the Asia-Oceania Federation of Organizations for Medical Physics (AFOMP), the South East Asia Federation of Organizations for Medical Physics (SEAFOMP), the Malaysian Institute of Physics (IFM) and a founding member of the ASEAN College of Medical Physics (ACOMP). Dr. Yeong has published more than 36 peer-reviewed journal papers, one academic book, 2 book chapters, 10 proceedings and more than 80 scientific papers. Her research interests focus on theranostics, image-guided minimally invasive cancer therapies, nanotherapeutics, 3D printing and radiation protection in medicine. She is currently leading the Cancer Innovation and Metabolic research group at Taylor's University.

Alan Christopher Perkins, PhD, FIPEM, HonFRCP, is a clinical professor of medical physics in the School of Medicine at the University of Nottingham and honorary consultant clinical scientist at Nottingham University Hospitals NHS Trust where he is a divisional lead for research and innovation. He has had over 35 years' experience in nuclear medicine and medical physics and broad managerial experience in the NHS. He has undertaken extensive research and development work with clinical, academic and industrial collaborators in nuclear medicine, gastroenterology, radiopharmacology, drug delivery and radiation protection. His contribution to this work has resulted in authorship of over 200 peer-reviewed publications and 6 published books. Professor Perkins is a past president of the British Nuclear Medicine Society and the International Research Group on Immuno-scintigraphy and Therapy, a previous vice president of the Institute of Physics and Engineering in Medicine and currently a governor and chair of the Research Strategy Board for Coeliac UK. He is an editor of the UK journal *Nuclear Medicine Communications*, and for over nine years has represented the UK on the High-Level Group for the Security of Medical Radioisotope Supplies at the Organisation for Economic Co-operation and Development (OECD). He has consulted for a number of commercial organisations and has acted as an expert witness for pharmaceutical litigation in the United States.

Acknowledgements

We acknowledge the contribution from the following people:

Azlan Che Ahmad
Department of Biomedical Imaging
Faculty of Medicine
University of Malaya
Kuala Lumpur, Malaysia

Muhammad Shahrun Nizam Daman Huri
Department of Biomedical Imaging
Faculty of Medicine
University of Malaya
Kuala Lumpur, Malaysia

David Lurie
Bio-Medical Physics
School of Medicine, Medical
    Sciences & Nutrition
University of Aberdeen
Aberdeen, United Kingdom

Juergen Meyer
Radiation Oncology Department
University of Washington Medical
    Center
Washington, District of Columbia

Khadijah Ramli
Department of Biomedical Imaging
Faculty of Medicine
University of Malaya
Kuala Lumpur, Malaysia

Mohammad Nazri Md Shah
Department of Biomedical Imaging
Faculty of Medicine
University of Malaya
Kuala Lumpur, Malaysia

Li Kuo Tan
Department of Biomedical Imaging
Faculty of Medicine
University of Malaya
Kuala Lumpur, Malaysia

Ngie Min Ung
Clinical Oncology Unit
Faculty of Medicine
University of Malaya
Kuala Lumpur, Malaysia

Wil van der Putten
Department of Medical Physics and
    Bioengineering
Galway University Hospitals
Galway, Ireland

Jeannie Hsiu-Ding Wong
Department of Biomedical Imaging
Faculty of Medicine
University of Malaya
Kuala Lumpur, Malaysia

List of abbreviations

A	mass number
ACD	annihilation coincidence detection
C-11	carbon-11
CFOV	central field-of-view
COR	centre of rotation
Cr-51	chromium-51
CT	computed tomography
CZT	cadmium zinc telluride
ECG	electrocardiogram
ELISA	enzyme-linked immunosorbent assay
F-18	fluorine-18
FDA	Food and Drug Administration
FDG	fluorodeoxyglucose
FOV	field-of-view
FWHM	full width at half maximum
He	helium
HVL	half value layer
I-125	iodine-125
I-131	iodine-131
IAEA	International Atomic Energy Agency
ICRP	International Commission on Radiological Protection
In-111	indium-111
LEHR	low-energy high-resolution
LOR	line-of-response
LSF	line spread function
mAbs	monoclonal antibodies
MIRD	medical internal radiation dosimetry
Mo-99	molebdenum-99
MRI	magnetic resonance imaging
MTF	modulation transfer function
MUGA	multiple-gated acquisition
NEMA	National Electrical Manufacturers Association

N-13	nitrogen-13
NaI	sodium iodide
NaI(Tl)	sodium iodide doped with thallium
O-15	oxygen-15
O-18	oxygen-18
P-32	phosphorus-32
Pb	lead
PET	positron emission tomography
PHA	pulse height analyser
PMT	photomultiplier tube
Po-210	polonium-210
PSF	point spread function
PSPMT	position sensitive photomultiplier tube
QC	quality control
Ra-223	radium-223
Re-188	rhenium-188
RIA	radioimmunoassay
RIT	radioimmunotherapy
ROI	region-of-interest
RPO	radiation protection officer
Sm-153	samarium-153
SPECT	single photon emission computed tomography
SUV	standardised uptake value
$t_{1/2}$	half-life
T_e	effective half-life
T_b	biological half-life
T_p	physical half-life
Tc-99m	technetium-99m
TOF	time-of-flight
TVL	tenth value layer
U-235	uranium-235
UFOV	useful field-of-view
Y-90	yittrium-90
Z	atomic number

List of physical constants

Avogadro's number	$N_a = 6.022 \times 10^{23}$ atoms·mol^{-1}
Speed of light in vacuum	$c = 2.998 \times 10^8$ m/s $\approx 3 \times 10^8$ m·s^{-1}
Electron charge	$e = 1.602 \times 10^{-19}$ C
Electron and positron rest mass	$m_e = 9.109 \times 10^{-31}$ kg or 0.511 Me·V·c^{-2}
Proton rest mass	$m_p = 1.673 \times 10^{-27}$ kg or 938.3 Me·V·c^{-2}
Neutron rest mass	$m_n = 1.675 \times 10^{-27}$ or 939.6 Me·V·c^{-2}
Planck's constant	$h = 6.626 \times 10^{-34}$ J·s
Electric constant (dielectric permittivity of vacuum)	$\varepsilon_0 = 8.854 \times 10^{-12}$ c·V^{-1}·m^{-1}
Magnetic constant (permeability of vacuum)	$\mu_0 = 4\pi \times 10^{-7}$ V·s·a^{-1}·m^{-1}
Newtonian gravitation constant	$G = 6.672 \times 10^{-11}$ m^3·kg^{-1}·s^{-2}
Proton mass/electron mass	$m_p/m_e = 1836.0$
Specific charge of electron	$e/m_e = 1.758 \times 10^{11}$ c·kg^{-1}

Radioactivity and Nuclear Transformation

1

1.1 NUCLEAR STABILITY CURVE

PROBLEM

Figure 1.1 shows the nuclear stability curve. Name the types of radioactive decays that tend to happen at regions A, B and C.

Solution

 A: Beta decay
 B: Positron emission or electron capture
 C: Alpha decay

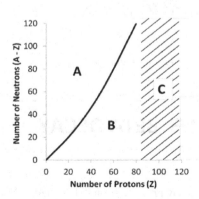

FIGURE 1.1 Nuclear stability curve.

1.2 ALPHA DECAY

PROBLEM

Explain the radioactive decay process through alpha decay. Your answer should include the physical principle, mechanism and decay equation.

Solution

Alpha decay is the spontaneous emission of an alpha particle (^4_2He or α) from an unstable nucleus (Figure 1.2). It typically occurs for heavy nuclides ($A > 150$ or $Z > 83$) when the nuclear binding energy can no longer hold the nucleons together. Alpha decay is often followed by gamma or characteristic X-ray emission resulting from the competing processes, such as internal conversion and Auger electron emission. An alpha particle has two protons and two neutrons and carries an electronic charge of 2+. Alpha decay can be expressed by the following equation:

$$^A_Z X \rightarrow {}^{A-4}_{Z-2}Y + {}^4_2He + transition\ energy$$

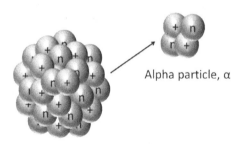

Alpha particle, α

Unstable heavy nuclei

FIGURE 1.2   Graphical illustration of alpha decay.

1.3 BETA DECAY

PROBLEM

Explain the radioactive decay process through beta decay. Your answer should include the physical principle, mechanism and decay equation.

Solution

When a nucleus has an excessive number of neutrons compared to protons (i.e. high N/Z ratio), the excess neutron will be converted into a proton, an electron and an antineutrino ($\overline{\nu_e}$). The proton remains in the nucleus, but the electron is emitted as a beta (β^-) particle (Figure 1.3). Beta decay increases the number of protons by 1 and hence transforms the atom into a different element with atomic number Z + 1. The mass number remains unchanged because of simultaneous decrease of a neutron only. Beta decay can be expressed by the following equation:

$$_Z^A X \rightarrow _{Z+1}^A Y + \beta^- + \overline{\nu_e} + energy$$

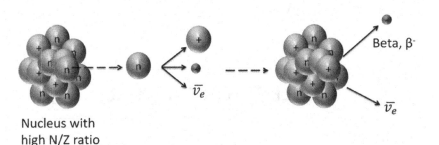

Nucleus with
high N/Z ratio

FIGURE 1.3   Graphical illustration of beta decay.

1.4 POSITRON DECAY

PROBLEM

Explain the radioactive decay process through positron decay. Your answer should include the principle, mechanism and decay equation.

Solution

When a nucleus has excess of protons compared to neutrons (i.e. low N/Z ratio), the excessive proton will be converted into a neutron, a positron (β^+), and a neutrino (ν) (Figure 1.4). The neutron will remain in the nucleus, but a positron will be emitted. Positron decay reduces the number of protons by 1 and hence transforms the atom into a different element with atomic number Z–1. The mass number remains unchanged because of the simultaneous increment of a neutron only.

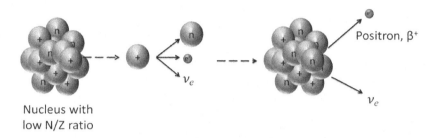

FIGURE 1.4   Graphical illustration of positron decay.

Positron decay can be expressed by the following equation:

$$_Z^A X \rightarrow \,_{Z-1}^A Y + \beta^+ + \nu_e + energy$$

1.5 ELECTRON CAPTURE

PROBLEM

Explain the radioactive decay process through electron capture. Your answer should include the principle, mechanism and decay equation.

Solution

Electron capture is an alternative process to positron decay when the nucleus has excessive number of protons compared to neutrons (i.e. low N/Z ratio). During electron capture, the nucleus captures an inner orbital (i.e. K- or L-shell) electron, thereby changing a proton to a neutron with the simultaneous emission of a neutrino (ν_e). The capture of an electron from the inner shell creates a vacancy in the orbit, which is then filled by an outer shell electron, and the energy difference between the electron shells results in the emission of a characteristic X-ray (Figure 1.5).

Electron capture can be expressed by the following equation:

$$_Z^A X + e^- \rightarrow \,_{Z-1}^A Y + \nu_e + energy$$

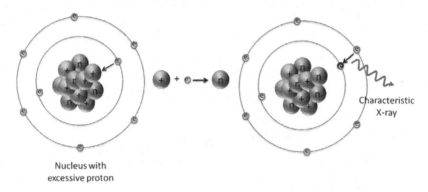

Nucleus with
excessive proton

FIGURE 1.5   Graphical illustration of electron capture.

1.6 ISOMERIC TRANSITION

PROBLEM

Explain the radioactive decay process through isomeric transition. Your answer should include the principle, mechanism and decay equation.

Solution

During radioactive decay, a daughter is often formed in an excited (metastable or isomeric) state. Gamma rays are emitted when the daughter nucleus undergoes an internal rearrangement and transitions from the excited state to a lower energy state (Figure 1.6). This process is called isomeric transition. Therefore, isomeric transition is a decay process that yields gamma radiation without the emission or capture of a particle by the nucleus. There is no change in atomic number (Z), mass number (A), or neutron number (N). Thus, this decay mode is isobaric and isotonic, and it occurs between two nuclear energy states with no change in the neutron-to-proton (N/Z) ratio. Isomeric transition can be expressed by the following equation:

$$_Z^A X^* \rightarrow {}_Z^A X + Energy$$

Note: An example of isomeric transition is the decay of Tc-99m.

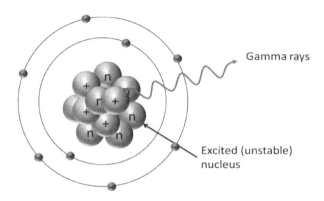

FIGURE 1.6   Graphical illustration of isomeric transition.

1.7 RADIATION PENETRABILITY

PROBLEM

Discuss the penetrability of alpha, beta, gamma and neutron radiation in matters, and suggest the suitable shielding materials for these radiations.

Solution

An alpha particle has the least penetrability, followed by beta and gamma radiation. Alpha can be stopped by a sheet of paper or skin, beta by aluminium, and gamma radiation by very thick concrete or lead. The penetrability of a neutron is dependent on its energy. Neutrons may penetrate through concrete and lead but can be slowed down by water or low atomic number materials, such as boron.

1.8 CALCULATION: NUMBER OF ATOMS

PROBLEM

Calculate the total number of atoms and total mass present in 370 MBq of Po-210, given the physical half-life of Po-210 is 138.4 days.

Solution

$$\lambda = \frac{ln2}{138.4\,\mathrm{d} \times 24\,\mathrm{h} \times 60\,\mathrm{min} \times 60\,\mathrm{s}} = 5.795 \times 10^{-8}\ \mathrm{s}^{-1}$$

$$A = 370\ \mathrm{MBq} = 3.7 \times 10^8\ \mathrm{Bq} = 3.7 \times 10^8\ \mathrm{dps}$$

- The activity, A, is related to the number of radioactive atoms as the following:

$$A = \lambda N$$

- Hence, $N = \dfrac{A}{\lambda} = \dfrac{3.7 \times 10^8\ \mathrm{dps}}{5.795 \times 10^{-8}\ \mathrm{s}^{-1}} = 6.38 \times 10^{15}$ atoms
- Total mass of Po-210 in 370 MBq $= \dfrac{N}{N_A}$
- Where N_A is the Avogadro's number, $N_A = 6.02 \times 10^{23}$
- Therefore, the total mass of Po-210 $= \dfrac{6.38 \times 10^{15} \times 210\ \mathrm{g}}{6.02 \times 10^{23}} = 2.23 \times 10^{-6}\ \mathrm{g}$
- Hence, 370 MBq Po-210 contains 6.38×10^{15} atoms and the total mass is 2.23 μg.

1.9 CALCULATION: SAMPLE COUNT RATE

PROBLEM

During radioactive counting, a 10-minute counting period of a sample plus background yields 10000 counts, and a 6-minute count of background alone yields 1296 counts. What are the net sample count rate and its standard deviation?

Solution

- Sample plus background count rate, $A = \dfrac{10000\ \mathrm{counts}}{10\ \mathrm{min}} = 1000\ \mathrm{cpm}$

- Background count rate, $B = \dfrac{1296\ \mathrm{counts}}{6\ \mathrm{min}} = 216\ \mathrm{cpm}$

- Net sample count rate $= 1000 - 216\ \mathrm{cpm} = 784\ \mathrm{cpm}$

- Standard deviation for $A = \sqrt{\dfrac{1000\ \mathrm{counts}}{10\ \mathrm{min}}} = 10\ \mathrm{cpm}$

- Standard deviation for $B = \sqrt{\dfrac{216 \text{ counts}}{6 \text{ min}}} = 6 \text{ cpm}$

- Therefore, net sample count rate $= 784 \pm \sqrt{10^2 + 6^2}$ cpm $= 784 \pm 12$ cpm

1.10 CALCULATION: THYROID UPTAKE

PROBLEM

The following data were obtained from the thyroid uptake study of a patient after administration of I-131 sodium iodide. Calculate the percentage and standard deviation of thyroid uptake in this patient.

REGION	COUNTS	TIME (MIN)
Standard	85500	2
Room background	952	2
Thyroid	42100	2
Thigh	2200	2

Solution

- Net standard count $= 85500 - 952 = 84548$
- Standard deviation for the standard counts, $\sigma_s = \sqrt{85500 + 952} = 294$
- Net thyroid count $= 42100 - 2200 = 39900$
- Standard deviation for the thyroid counts, $\sigma_t = \sqrt{42100 + 2200} = 210$
- % thyroid uptake $= \dfrac{39900}{84548} \times 100\% = 47.2\%$

- Standard deviation of the uptake $= \dfrac{39900}{84548} \times \sqrt{\left(\dfrac{294}{84548}\right)^2 + \left(\dfrac{210}{39900}\right)^2}$

$$= 0.472 \times \sqrt{0.00001209 + 0.00002770}$$

$$= 0.472 \times 0.006308$$

$$= 0.003$$

- % standard deviation of the uptake $= \dfrac{0.003}{0.472} \times 100\% = 0.6\%$

- Therefore, the percentage of I-131 uptake in the thyroid of the patient is $47.2 \pm 0.6\%$.

1.11 PHYSICAL HALF-LIFE (I)

PROBLEM

If a radionuclide decays at a rate of 25% per hour, what is its physical half-life?

Solution

$$A = A_o e^{-\lambda t}$$

The radionuclide decays from 100% to 75% after one hour, hence:

$$0.75 = 1.00 e^{-\lambda(1h)}$$

$$\ln(0.75) = -\lambda(1h)$$

$$\lambda = 0.288 \, h^{-1}$$

$$\lambda = \frac{\ln 2}{t_{1/2}}$$

Therefore, the physical half-life, $t_{1/2} = \dfrac{0.693}{0.288 \, h^{-1}} = 2.4h$

1.12 PHYSICAL HALF-LIFE (II)

PROBLEM

An unknown radioactive source found in a nuclear medicine department decays to 1/16 of its original activity after 24 hours. What is the physical half-life of this radioactive source? Based on your answer, suggest a possible name of the radioactive source.

Solution

$\dfrac{1}{16}$ activity equals to four half-lives.

$$\frac{24 \, h}{4 \text{ half-lives}} = 6h$$

The physical half-life of the source is six hours. The source is most probably Tc-99m, which is the most common radionuclide used in nuclear medicine.

1.13 EFFECTIVE HALF-LIFE (I)

PROBLEM

   a. Explain what is effective half-life.

   b. If Tc-99m with a physical half-life of six hours is used to radio-label a compound that clears from the body with a biological half-life of four hours, what is the effective half-life of the radiopharmaceutical?

Solution

   a. The effective half-life is defined as the time interval required to reduce the radioactivity level of an internal organ or the whole body to one-half of its original activity due to radioactive decay (physical half-life) and biological elimination (biological half-life). Effective half-life, T_e, is expressed as the following:

$$\frac{1}{T_e} = \frac{1}{T_p} + \frac{1}{T_b}$$

or

$$T_e = \frac{T_p \times T_b}{T_p + T_b}$$

where:

T_p = physical half-life, which is defined as the time interval required for a radioactive substance to decay to half of its original activity.

T_b = biological half-life, which is defined as the time interval required for an organ or the whole body to eliminate 50% of its original activity by normal routes of elimination: metabolic turnover and excretion.

   b. $T_e = \dfrac{T_p \times T_b}{T_p + T_b}$

Therefore, $T_e = \dfrac{6\,\text{h} \times 4\,\text{h}}{6\,\text{h} + 4\,\text{h}}$

$$T_e = 2.4\,\text{h}$$

Note: Effective half-life is always smaller than the biological half-life and physical half-life. For a permanently implanted radionuclide (which no biological half-life is applied), the effective half-life is the same as the physical half-life.

1.14 EFFECTIVE HALF-LIFE (II)

PROBLEM

Consider 10% of Tc-99m-DTPA is eliminated from the body of a patient by renal excretion, 25% by faecal excretion and 3% by perspiration within six hours. What is the effective half-life of the radiopharmaceutical, given that the physical half-life, T_p, for Tc-99m is six hours?

Solution

- Total biological elimination $= 10 + 25 + 3 = 38\%$
- Given that 38% elimination in six hours.
- To achieve 50% elimination (biological half-life, T_b), it takes:

$$T_b = \frac{50}{38} \times 6\,h = 7.89\ h$$

$$T_e = \frac{T_b \times T_p}{T_b + T_p} = \frac{7.89\,h \times 6\,h}{7.89 + 6\,h} = 3.4\,h$$

- Therefore, the effective half-life, $T_e = 3.4\ h$

1.15 RADIOACTIVE DECAY EQUATION

PROBLEM

Derive the radioactive decay formula, starting from

$$A = -\lambda N$$

where:
   A = activity (Bq)
   λ = decay constant (s^{-1})
   N = number of particles

Solution

Given:

$$A = -\lambda N = \frac{dN}{dt}$$

$$-\frac{dN}{N} = \lambda dt$$

By applying integration to the formula above:

$$-\int_{N_0}^{N_t} \frac{dN}{N} = \lambda \int_0^t dt$$

$$ln\left(\frac{N_t}{N_0}\right) = -\lambda t$$

$$\frac{N_t}{N_0} = e^{(-\lambda t)}$$

$$N_t = N_0 e^{(-\lambda t)}$$

1.16 RADIOACTIVE DECAY CALCULATION (I)

PROBLEM

A patient is injected with 1110 MBq Tc-99m-MDP ($T_{p1/2} = 6.02$ hours) for a whole-body bone scan. Two hours later, the patient is imaged with a gamma camera. Assuming that 20% of the activity has been excreted by the kidneys, how much activity remains at the time of imaging?

Solution

Given:

$$A_0 = 1110\,\text{MBq}$$

$$T_{p1/2} = 6.02\,\text{h}$$

$$T_{b1/2} = \frac{50\%}{20\%} \times 2\,\text{h} = 5\,\text{h}$$

$$T_e = \frac{T_p \times T_b}{T_p + T_b} = \frac{6.02 \times 5}{6.02 + 5} = 2.73\,\text{h}$$

$$\lambda = \frac{\ln 2}{T_e} = \frac{\ln 2}{2.73\,\text{h}} = 0.254\,\text{h}^{-1}$$

$$A = A_0 e^{-\lambda t}$$

$$= 1110\,\text{MBq}\, e^{-0.254\,\text{h}^{-1}(2\text{h})}$$

$$= 667.88\,\text{MBq}$$

Therefore, the remaining activity at the time of imaging is approximately 668 MBq.

1.17 RADIOACTIVE DECAY CALCULATION (II)

PROBLEM

A thyroid cancer patient received an oral administration of 3700 MBq radioiodine (I-131), and 50% of the I-131 was excreted from the body after 48 hours. What is the remaining activity in the patient's body on the third day of the treatment? Assume the physical half-life for I-131 is eight days.

Solution

$$T_e = \frac{T_p \times T_b}{T_p + T_b} = \frac{8\,\text{d} \times 2\,\text{d}}{8\,\text{d} + 2\,\text{d}} = 1.6\,\text{d}$$

$$\lambda = \frac{\ln 2}{T_e} = \frac{\ln 2}{1.6\,\text{d}} = 0.433\,\text{d}^{-1}$$

$$A = A_0 e^{-\lambda t}$$

$$= 3700\,\text{MBq}\; e^{-0.433\ \text{d}^{-1}(3\,\text{d})}$$

$$= 1009\,\text{MBq}$$

Therefore, the remaining activity in patient's body on the third day of the treatment is approximately 1009 MBq.

1.18 RADIOACTIVE DECAY CALCULATION (III)

PROBLEM

How long does it take for a sample of 370 MBq I-123 ($t_{1/2} = 13.2$ h) and a sample of 1850 MBq Tc-99m ($t_{1/2} = 6$ h) to reach the same activity?

Solution

$$\lambda = \frac{ln\,2}{t_{1/2}}$$

$$\lambda_1 = \frac{ln\,2}{13.2\,\text{h}} = 0.0525\ \text{h}^{-1}$$

$$\lambda_2 = \frac{ln\,2}{6\,\text{h}} = 0.1155\ \text{h}^{-1}$$

$$A_1 e^{-\lambda_1 t} = A_2 e^{-\lambda_2 t}$$

$$370 e^{-0.0525 t} = 1850 e^{-0.1155 t}$$

$$\frac{e^{-0.0525 t}}{e^{-0.1155 t}} = \frac{1850\,\text{MBq}}{370\,\text{MBq}}$$

$$e^{(-0.0525 t)-(-0.1155 t)} = 5$$

$$e^{0.063 t} = 5$$

$$0.063 t = 1.609$$

$$t = 25.5\ \text{h}$$

Therefore, 370 MBq I-123 and 1850 MBq Tc-99m will achieve the same activity at 25.5 hours later.

1.19 ATTENUATION

PROBLEM

The mass attenuation coefficient of bone with a density of 1.8 g·cm⁻³ is 0.2 cm²g⁻¹ for an 80 keV gamma ray. Calculate the percentage of a photon beam attenuated by 5 cm thickness of bone.

Solution

Mass attenuation coefficient $= \dfrac{\mu}{\rho}$

$$\frac{\mu}{1.8\,\mathrm{g}\cdot\mathrm{cm}^{-3}} = 0.2 \ \mathrm{cm^2 g^{-1}}$$

$$\mu = 0.36 \ \mathrm{cm^{-1}}$$

Fraction transmitted:

$$\frac{N_t}{N_0} = e^{-\mu x}$$

$$\frac{N_t}{N_0} = e^{-0.36\,\mathrm{cm}^{-1}(5\,\mathrm{cm})} = 0.165$$

Percentage attenuated $= (1 - 0.165) \times 100\% = 83.5\%$

1.20 GAMMA RAY CONSTANT

PROBLEM

a. What is the specific gamma ray constant (Γ)?
b. Why should a radionuclide for imaging have a high specific gamma ray constant?

Solution

a. The specific gamma ray constant (Γ) is the exposure rate at a specific distance from a given amount of a photon-emitting radionuclide. These constants are used frequently for dosimetry and radiation protection purposes. The SI unit for Γ is $Ckg^{-1}s^{-1}Bq^{-1}$ at 1 m, but it is often expressed in its traditional units, $R\ h^{-1}mCi^{-1}$ at 1 cm.

b. A high Γ means there are large numbers of gamma rays emitted and available to form the image. Some radionuclides decay by more than one mechanism, so the radionuclides that produce more useful gamma rays are preferable for nuclear medicine imaging.

1.21 ALPHA PARTICLE RANGE

PROBLEM

Calculate the range of a 6.3 MeV alpha particle in:

a. Air (density = 0.001293 $g \cdot cm^{-3}$, atomic mass number ~ 14.8)
b. Plastic (density = 1.13 $g \cdot cm^{-3}$, atomic mass number ~ 11.81)
c. Mylar (density = 1.4 $g \cdot cm^{-3}$, atomic mass number ~ 12.88)
d. Human tissue (density = 1.04 $g \cdot cm^{-3}$, atomic mass number ~ 13.6)

Solution

a. To calculate alpha particle range in air for $4 < E < 8$ MeV, the following equation is used:

R (cm) = 0.325 $E^{3/2}$ (MeV)

$R = 0.325(6.3)^{3/2}$

$R = 5.14$ cm

b. To calculate the alpha particle range in other material, the scaling law is used:

$$\frac{R_1}{R_0} = \frac{\rho_0 \sqrt{A_1}}{\rho_1 \sqrt{A_0}}$$

$$\frac{R_{plastic}}{R_{air}} = \frac{\rho_{air} \sqrt{A_{plastic}}}{\rho_{plastic} \sqrt{A_{air}}}$$

$$R_{\text{plastic}} = \frac{0.001293\sqrt{11.81}}{1.13\sqrt{14.8}}(5.14 \text{ cm})$$

$$R_{\text{plastic}} = \frac{0.00444}{4.3472}(5.14 \text{ cm})$$

$$R_{\text{plastic}} = 0.0053 \text{ cm}$$

c. $$\frac{R_{\text{mylar}}}{R_{\text{air}}} = \frac{\rho_{\text{air}}\sqrt{A_{\text{mylar}}}}{\rho_{\text{mylar}}\sqrt{A_{\text{air}}}}$$

$$R_{\text{mylar}} = \frac{0.001293\sqrt{12.88}}{1.4\sqrt{14.8}}(5.14 \text{ cm})$$

$$R_{\text{mylar}} = \frac{0.00464}{5.3859}(5.14 \text{ cm})$$

$$R_{\text{mylar}} = 0.0044 \text{ cm}$$

d. $$\frac{R_{\text{human tissue}}}{R_{\text{air}}} = \frac{\rho_{\text{air}}\sqrt{A_{\text{human tissue}}}}{\rho_{\text{human tissue}}\sqrt{A_{\text{air}}}}$$

$$R_{\text{human tissue}} = \frac{0.001293\sqrt{13.6}}{1.04\sqrt{14.8}}(5.14 \text{ cm})$$

$$R_{\text{human tissue}} = \frac{0.00477}{4.0009}(5.14 \text{ cm})$$

$$R_{\text{human tissue}} = 0.0061 \text{ cm}$$

Radionuclide Production and Radiopharmaceuticals

2

2.1 CHARACTERISTICS OF IDEAL RADIOPHARMACEUTICAL FOR DIAGNOSTIC NUCLEAR MEDICINE

PROBLEM

List eight criteria of an ideal radiopharmaceutical used for diagnostic nuclear medicine imaging studies.

Solution

The criteria of an ideal radiopharmaceutical for diagnostic nuclear medicine include:

- Low radiation dose to patient.
- Appropriate physical half-life for the duration of the studies.
- A pure gamma emitter.
- Appropriate energy for imaging purposes (about 100 to 200 keV).
- Decays to a stable nuclide.
- High target to non-target uptake of activity.
- Good chemical reactivity (ability to form stable complexes with carrier molecules).
- Cost-effective and readily available.

2.2 CHARACTERISTICS OF IDEAL RADIOPHARMACEUTICAL FOR THERAPEUTIC NUCLEAR MEDICINE

PROBLEM

List seven criteria of an ideal radiopharmaceutical used for therapeutic nuclear medicine.

Solution

The criteria of an ideal radiopharmaceutical for therapeutic nuclear medicine include:

- Moderately long, effective half-life (measured in days).
- High linear energy transfer, such as α- and β-particles.
- Emission of high-radiation energy, for example, >1 MeV.
- High target-to-non-target ratio to minimize radiation dose to non-target tissues.
- Decays into stable daughter products.
- Minimal radiation exposure to personnel in contact with patient.
- Cost-effective and readily available.

2.3 PHYSICAL PROPERTIES AND DECAY SCHEME OF Tc-99m

PROBLEM

a. Describe the physical properties of Tc-99m.
b. Sketch and label the radioactive decay scheme of Tc-99m.

Solution

a. Tc-99m is a colourless metallic chemical element. It has an atomic number, Z, of 43, and mass number, A, of 56. It is metastable (as indicated by the symbol 'm') and thereby decays into Tc-99 through gamma emission (88%) and isomeric transition (12%). It has a physical half-life of 6.02 hours and emits 140 keV gamma rays, which are readily detectable by a gamma camera. Tc-99m has multiple valence states. There are 21 known isotopes and numerous isomers from the family of technetium whereby all of them are radioactive.

b. The radioactive decay scheme of Tc-99m is illustrated in Figure 2.1.

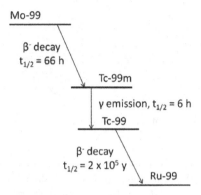

FIGURE 2.1   Radioactive decay scheme of Tc-99m.

2.4 CYCLOTRON

PROBLEM

With the aid of a diagram, explain the basic principles of radionuclide production in a cyclotron. Use In-111 as an example.

Solution

Figure 2.2 shows the basic structure of a cyclotron. A cyclotron consists of two large dipole magnets (known as Dees) that produce a uniform magnetic field. An oscillating voltage is applied to produce an electric field across the gap between the Dees. Charged particles, such as protons and electrons, are injected into the magnetic field region of the Dees. The electric field in the gap then accelerates the particles as they pass through it. The particles now have higher energy so they follow a semi-circular path in the next Dee with a larger radius and then reach the gap again. The field in the gap accelerates them, and they enter the first Dee again. Thus, the particles gain energy as they spiral outwards until they gain sufficient velocity and are deflected into a target.

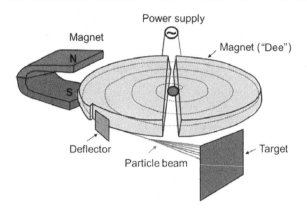

FIGURE 2.2 Basic structure of a cyclotron.

In the production of In-111, the accelerated particles are protons. The target atoms are Cd-111. When a proton enters the nucleus of a Cd-111 atom, the Cd-111 is transformed into In-111 by discharging a neutron. This interaction in the form of nuclear reaction is as follows:

Target atom (bombarding particle, emitted particle) product nuclide
For example:

Cd-111 (proton, neutron) In-111

2.5 CYCLOTRON-PRODUCED RADIONUCLIDES

PROBLEM

Name three radionuclides other than In-111 that are produced by a cyclotron. State their nuclear reaction equations and physical half-lives.

Solution

RADIONUCLIDE	NUCLEAR REACTION	PHYSICAL HALF-LIFE
Fluorine-18	O-18 (p, n) F-18	110 min
Nitrogen-13	C-12 (d, n) N-13	10 min
Oxygen-15	N-14 (d, n) O-15 or N-15 (p, n) O-15	2 min

2.6 NUCLEAR FISSION

PROBLEM

- Define nuclear fission.
- Explain the fission mechanism in a nuclear reactor using Uranium-235 (U-235) as the fuel.

Solution

Nuclear fission is the splitting of a heavy nuclide into daughter products (smaller atoms) with the release of energy (Figure 2.3).

- In a nuclear reactor, fission is triggered by a neutron splitting the nucleus of a heavy atom, such as U-235.
- When U-235 absorbs a neutron, the resulting nucleus (U-236) is in an extremely unstable excited energy state that usually promptly splits into two smaller nuclei, known as the fission fragments.
- The fission fragments travel with very high kinetic energies, with the simultaneous release of gamma radiation and two to five free neutrons (the number of ejected neutrons depends on how the U-235 atom splits).
- Some neutrons are absorbed by non-fissionable material in the reactor, whereas others are moderated and absorbed by U-235 atoms and induce additional fissions. This is known as the nuclear chain reaction.
- The ratio of the number of fissions in one generation to the number of fissions in the previous generation is called the multiplication factor.
- When the multiplication factor = 1, the number of fissions per generation is constant and the reactor is said to be critical.
- When the multiplication factor > 1, the rate of the chain reaction increases and the reactor is said to be supercritical.
- When the multiplication factor < 1, the rate of the chain reaction decreases (more neutrons being absorbed than produced) and the reactor is said to be subcritical.
- The energy released by the nuclear fission of a U-235 atom is approximately 200 MeV.

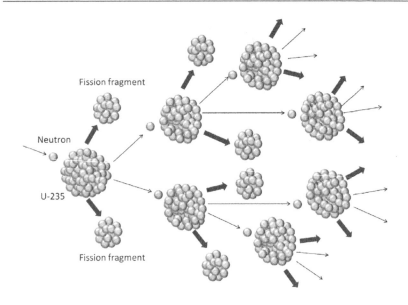

FIGURE 2.3   The process of nuclear fission.

2.7 REACTOR-PRODUCED RADIONUCLIDES

PROBLEM

- Name three radionuclides other than Mo-99 that are produced by a nuclear reactor.
- State their nuclear reaction equations and physical half-lives.

Solution

RADIONUCLIDE	REACTION	PHYSICAL HALF-LIFE
Phophorus-32	$^{31}P(n, \gamma)^{32}P$	14.3 days
Chromium-51	$^{50}Cr(n, \gamma)^{51}Cr$	27.8 days
Iodine-125	$^{124}Xe(n, \gamma)^{125}Xe \xrightarrow{EC \ or \ \beta^{+}} {}^{125}I$	59.4 days

2.8 Mo-99/Tc-99m
RADIONUCLIDE GENERATOR (I)

PROBLEM

Sketch a labelled diagram of a Mo-99/Tc-99m radionuclide generator and describe the process of Tc-99m elution.

Solution

Figure 2.4 shows the basic structure of a Mo99/Tc-99m generator. The generator is heavily shielded in a lead container known as the 'lead pig'. Chemically purified Mo-99 is loaded in the form of ammonium molybdenate ($NH_4^+ MoO_4^-$) onto a porous column containing 5–10 g of alumina (Al_2O_3) resin. The ammonium molybdenate is attached to the surface of the alumina

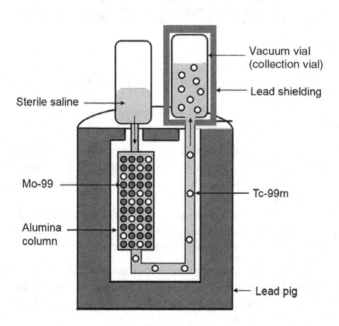

FIGURE 2.4 Block diagram of a Mo-99/Tc-99m generator.

molecules through adsorption. The column is adjusted to an acid pH to promote binding. When the Mo-99 decays, it forms pertechnetate (TcO_4^-), which is less tightly bound to the alumina compared to Mo-99 because of its single charge. By passing normal saline solution (NaCl) over the column, Tc-99m is removed or washed off from the column and collected by negative pressure into a vacuum vial at the collecting port in the form of sodium pertechnetate ($NaTcO_4$). The vacuum vial is placed inside a lead shielding for radiation protection purposes.

Note: In some generator systems, the sterile saline solution is held in a reservoir. Elution of the required amount of product is controlled by the vacuum in the collection vial. Sterility is ensured by terminal filtration at the point just before the solution enters the collection vial.

2.9 Mo-99/Tc-99m
RADIONUCLIDE GENERATOR (II)

PROBLEM

Differentiate between the wet and dry systems of the Mo-99/Tc-99m generator.

Solution

The wet system has a saline reservoir connected by tubing to one end of the column, and an evacuated vial draws the saline through the column, hence keeping the column wet. During eluation, a vacuum vial is placed at the exit port to collect the Tc-99m eluent.

The dry system does not have a saline reservoir in the generator; therefore, it requires a small vial containing normal saline to be attached at the entry port and a vacuum vial at the collection port during eluation.

Note: Evacuated (vacuum) vials are used since this creates a condition of negative pressure during generator elution. This prevents the egress of solution and the possibility of area contamination should any parts of the generator or tubing develop a puncture or leak.

2.10 UNDESIRABLE Al³⁺ IN Tc-99m ELUTION

PROBLEM

a. Why is Mo-99 undesirable in the Tc-99m eluate from a generator?
b. Why is Al^{3+} undesirable in the Tc-99m eluate from a generator?

Solution

a. Mo-99 is the parent radionuclide of Tc-99m and is bound to the alumina column in the generator. It is essential that no Mo-99 separates from the column during elution, as this would both degrade the gamma camera image and contribute unnecessary additional radiation doses to the patient. Mo-99 has a physical half-life of 66 hours and emits high energy of 740 keV gamma rays.

b. Al^{3+} is a trace metal that should not be administered into the patients. It will interfere with the preparation of Tc-99m radiopharmaceuticals, such as Tc-99m-labelled colloid and Tc-99m bone-scanning agents by forming larger particles, which will be trapped in the small capillaries of the lungs following intravenous injection. It will also agglutinate the red blood cells during labelling.

Note: According to the FDA rules and regulations, the concentration of aluminium ion should not be more than 10 µg/mL of the generator eluate.

2.11 TRANSIENT EQUILIBRIUM

PROBLEM

If a parent radionuclide, p, decays to a daughter radionuclide, d, which in turn decays to another radionuclide, the rate of growth of d is given as:

$$\frac{dN_d}{dt} = \lambda_p N_p - \lambda_d N_d \tag{2.1}$$

By integrating the above equation, we get:

$$(A_d)_t = \lambda_d N_d = \frac{\lambda_d (A_p)_0}{(\lambda_d - \lambda_p)} (e^{-\lambda_p t} - e^{-\lambda_d t}) \qquad (2.2)$$

where:

N_p = number of particles of the parent radionuclide
N_d = number of particles of the daughter radionuclide
λ_p = decay constant for the parent radionuclide
λ_d = decay constant for the daughter radionuclide
A_p = activity of the parent radionuclide
A_d = activity of the daughter radionuclide
t = time

From Equations 2.1 and 2.2, derive an equation that shows transient equilibrium. What is the condition for transient equilibrium to happen?

Solution

If $\lambda_d > \lambda_p$, that is $(t_{1/2})_d < (t_{1/2})_p$, then $e^{-\lambda_d t}$ is negligible compared to $e^{-\lambda_p t}$ when t is sufficiently long.

The equation becomes:

$$(A_d)_t = \frac{\lambda_d (A_p)_0}{(\lambda_d - \lambda_p)} \left(e^{-\lambda_p t} \right)$$

$$= \frac{\lambda_d}{(\lambda_d - \lambda_p)} (A_p)_t$$

$$= \frac{\dfrac{0.693}{(t_{1/2})_p}}{\dfrac{0.693}{(t_{1/2})_p} - \dfrac{0.693}{(t_{1/2})_d}} (A_p)_t$$

$$= \frac{(t_{1/2})_p}{(t_{1/2})_p - (t_{1/2})_d} (A_p)_t$$

This equation is known as transient equilibrium. The equilibrium only holds good when $(t_{1/2})_p$ and $(t_{1/2})_d$ differ by a factor between 10 and 50.

2.12 Tc-99m TRANSIENT EQUILIBRIUM

PROBLEM

Figure 2.5 shows the activity-time plot of Mo-99 and Tc-99m decay.

    i. What does curve X indicate?
    ii. What does curve Y indicate?
    iii. Explain briefly the phenomenon demonstrated by the plot.

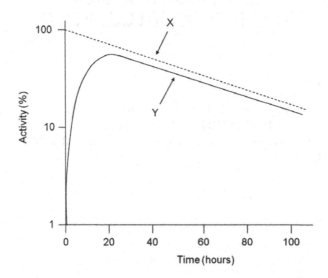

FIGURE 2.5   Activity-time plot of Mo-99 and Tc-99m.

Solution

    i. Curve X indicates the radioactive decay of the parent radionuclide, that is, Mo-99.
    ii. Curve Y indicates the radioactivity of the daughter (i.e. Tc-99m) at subsequent time intervals.
    iii. The plot demonstrates the phenomenon of transient equilibrium. The activity of Mo-99 (curve X) and Tc-99m (curve Y) versus time were plotted on a logarithm graph. The activity of the daughter, Tc-99m, initially builds up as a result of the decay of the parent,

reaches a maximum at approximately 23 hours, then achieves transient equilibrium whereby the daughter decays at the same rate as the parent. Generally, the daughter activity is higher than the parent activity at equilibrium. However, in Mo-99 decay, only 87.6% of Mo-99 decays to Tc-99m and the remaining 12.4% decays to Tc-99 directly; hence, in the time-activity plot, the Tc-99m activity is lower than the Mo-99.

2.13 Mo-99 TRANSIENT EQUILIBRIUM CALCULATION

PROBLEM

Mo-99 ($t_{1/2} = 66$ hours) and Tc-99m ($t_{1/2} = 6$ hours) are in transient equilibrium in a Mo generator. If 1000 MBq of Mo-99 is present in the generator, what would be the activity of Tc-99m after 82 hours, assuming that 87% of Mo-99 decays to Tc-99m?

Solution

$$(A_p)_t = (A_p)_0 e^{\left(-\frac{0.693}{66\,h}\right)(82\,h)}$$

$$= 1000\,\text{MBq}\, e^{(-0.861)}$$

$$= 422.74\,\text{MBq}$$

Activity of Tc-99m after 82 hours is:

$$(A_d)_t = \left(\frac{(t_{1/2})_p}{(t_{1/2})_p - (t_{1/2})_d}\right)(A_p)_t\left(0.87\right)$$

$$= \left(\frac{66}{60}\right)(422.74\ MBq)(0.87)$$

$$= 404.56\,\text{MBq}$$

2.14 SECULAR EQUILIBRIUM EQUATION

PROBLEM

As a continuation from Section 2.11, derive the equation that shows secular equilibrium. What is the condition for secular equilibrium to happen?

Solution

When λ_d is much greater than λ_p, that $(t_{1/2})_d$ is much shorter than $(t_{1/2})_p$, the λ_d can be neglected compared to λ_p. Equation 2.2 now becomes:

$$(A_d)_t = \frac{\lambda_d (A_p)_0}{\lambda_d}\left(e^{-\lambda_p t}\right)$$

$$(A_d)_t = (A_p)_0 \left(e^{-\lambda_p t}\right)$$

$$= (A_p)_t$$

The condition for secular equilibrium to happen is when the physical half-life of the parent is much longer than the daughter, by a factor of 100 or greater.

2.15 Mo-99 BREAKTHROUGH TEST

PROBLEM

With the aid of a diagram, describe briefly the quality control procedures of the Mo-99 breakthrough test.

Solution

Figure 2.6 illustrates the method for the Mo-99 breakthrough test. The vial containing the Tc-99m eluate is placed in a thick (~6 mm) lead container and then placed into a radionuclide calibrator. The high-energy photons of Mo-99 can be detected, whereas virtually all of the Tc-99m 140 keV photons are attenuated by the lead container. The acceptance limit for Mo-99 breakthrough is 5.55 kBq Mo-99 per 37 MBq of Tc-99m eluate.

Note: This procedure should be undertaken on the first elution from each new generator and at any time when it is suspected that the generator has not been used according to accepted normal standard procedures. It is also good practice to check for breakthrough at the end of use of the generator to confirm the integrity of the column.

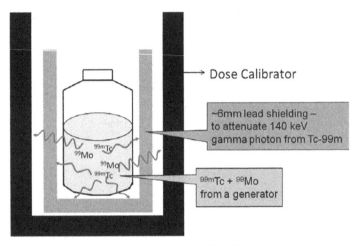

Acceptance limit: <0.15 µCi Mo-99 per 1 mCi of Tc-99m eluate

FIGURE 2.6 Mo-99 breakthrough test.

2.16 PREPARATION OF RADIOPHARMACEUTICAL

PROBLEM

A vial containing 5 mL of Tc-99m-MDP is prepared in the nuclear medicine hot-lab for radionuclide bone imaging. At 8.00 a.m., the calibrated activity of the radiopharmaceutical is 3700 MBq. What volume of the preparation is needed to give a 740 MBq injection at 12.00 p.m., given the physical half-life of Tc-99m is six hours?

Solution

Decay constant λ for Tc-99m:

$$\lambda = \frac{0.693}{t_{1/2}} = \frac{0.693}{6\,h} = 0.115\,h^{-1}$$

Using the decay equation,

$$N = N_o e^{-\lambda t}$$

$$N = 3700\,e^{-\left(0.115\,h^{-1}\right)\left(4\,h\right)} = 2335\,MBq$$

$$5\,mL \rightarrow 2335\,MBq$$

$$?\,mL \rightarrow 740\,MBq$$

$$\text{Volume of preparation} = \frac{740\,MBq}{2335\,MBq} \times 5\,mL = 1.58\,mL$$

2.17 ADMINISTRATION OF RADIOPHARMACEUTICALS

PROBLEM

A patient was given a radioactive meal labelled with 1110 MBq In-111 for a colonic transit study. The patient was imaged 24 hours later. Assuming that none of the activity was excreted, what is the remaining total activity at the time of imaging, given the physical half-life for In-111 is 67.9 hours?

Solution

$$A = A_o e^{-\lambda t}$$

$$A_o = 1110\,MBq$$

$$\lambda = \frac{0.693}{67.9\,h} = 0.01\,h^{-1}$$

$$A = 1110\,MBq\,e^{\left(-0.01\,h^{-1}\right)\left(24\,h\right)} = 873\,MBq$$

Note: Non-absorbable tracers are commonly used for gastric emptying and gastrointestinal transit.

2.18 PET RADIOPHARMACEUTICAL

PROBLEM

Name four radiopharmaceuticals used for positron emission tomography (PET) imaging, each containing a different positron emitting radionuclide, state the physical half-life in each case.

Solution

RADIOPHARMACEUTICAL	PHYSICAL HALF-LIFE (MIN)
F-18-FDG or fluorodeoxyglucose	110
O-15-carbon dioxide or O-15-carbon monoxide	2.05
C-11-acetate or C-11-carbon monoxide or C-11-carbon dioxide	20.4
N-13-ammonia	10

2.19 F-18-Fluorodeoxyglucose

PROBLEM

What is F-18-FDG? Explain the metabolism of F-18-FDG in human tissues that makes it a useful radiopharmaceutical in detecting and staging of malignant tumours.

Solution

F-18-FDG stands for fluorine-18-flurodeoxyglucose. It is a glucose analogue, with the positron emitting radionuclide, F-18 substituted for the normal hydroxyl group (-OH) at the C-2 position in the glucose molecule. Figure 2.7 shows the difference between the F-18-FDG and glucose molecule structures.

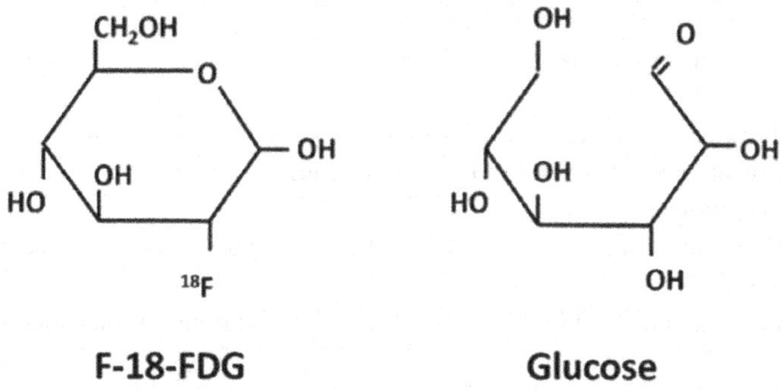

FIGURE 2.7 Comparison of the F-18-FDG and glucose molecule structures.

F-18-FDG follows the same transport route as glucose. After administration via intravenous injection, F-18-FDG is actively transported into the cells by specific glucose transporters. Once inside the cell, F-18-FDG is phosphorylated by hexokinase enzyme and forms F-18-FDG-6 phosphate. Normally, once phosphorylated, glucose continues along the glycolytic pathway for energy production. However, F-18-FDG cannot undergo glycolysis because the (−OH) group is missing. The F-18-FDG-6-phosphate molecules are therefore effectively trapped inside the cells. As a result, the distribution of F-18-FDG is a good reflection of the glucose uptake and phosphorylation by cells in the body. Malignant cells replace oxygen respiration by fermentation of sugar and so accumulate the F-18-FDG at a higher rate than normal cells. The uptake can be imaged and measured using a PET scanner.

2.20 QUALITY CONTROL OF RADIOPHARMACEUTICALS

PROBLEM

What are the factors that may induce radiochemical impurities in a radiopharmaceutical?

Solution

Radiochemical impurities may be due to:

- Temperature changes in storage and manufacturing areas.
- Presence of unwanted oxidizing or reducing agents.
- pH changes.
- Radiation damage to the chemical constituents of the radiopharmaceuticals.
- Poor standards of handling and lack of sterile preparation procedures.

Note: Radiopharmaceuticals should always be prepared in an appropriate radiopharmacy environment according to accepted standards with appropriate batch manufacturing recording and quality control measures in place.

Non-imaging Detectors and Counters

3

3.1 DEAD TIME

PROBLEM

Define 'dead time' of a counting system.

Solution

'Dead time' of a counting system refers to the period where the counter cannot process a second event after the first event has been detected. Typical values of dead time are between 250 and 1000 ns.

3.2 PARALYSABLE AND NON-PARALYSABLE COUNTING SYSTEMS

PROBLEM

With the aid of diagrams, explain the differences between paralysable and non-paralysable counting systems.

Solution

Figure 3.1 shows the differences between paralysable and non-paralysable counting systems. The black bars indicate processed signals, whereas the grey bars indicate non-processed signals. In non-paralysable systems, a fixed dead time, τ, follows each event that occurs during the live period of the detector. Signals that arrive during the dead time are not recorded and have no effect on the system (i.e. signals 3 and 5). However, in paralysable systems, in addition to the usual dead time effect (signals 3 and 5), the arrival of a secondary signal before complete processing of the previous signals will extend the dead time, causing the secondary signal to be rejected (i.e. signal 6).

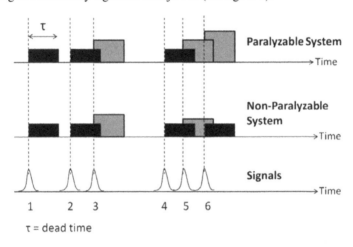

FIGURE 3.1 Paralysable and non-paralysable counting systems.

Note: In extreme cases, where the count rate is high, paralysable systems can be completely unresponsive (switched off) as the signals and dead times overlap and no signal is registered, hence the term 'paralysed'.

3.3 PARALYSABLE AND NON-PARALYSABLE COUNTING SYSTEMS: COUNT RATE RESPONSE

PROBLEM

Refer to Figure 3.2. If A is the theoretical response of an ideal detection system, name the detection systems of B and C. Explain your answers.

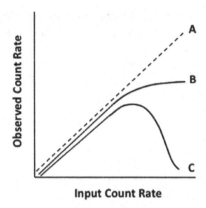

FIGURE 3.2   Observed count rate versus input count rate of different types of detection systems.

Solution

B is the non-paralysable system. The dead time increases with increasing input count rate and becomes saturated in the plateau region.

C is the paralysable system. As the input count rate increases beyond the saturation region, the system becomes paralysed and loses its detection efficiency rapidly beyond the saturation region.

3.4 BASIC PRINCIPLE OF GAS-FILLED DETECTORS

PROBLEM

With the aid of a diagram, explain briefly the principle of operation of a gas-filled detector.

Solution

Figure 3.3 illustrates the basic operational principle of a gas-filled detector. A gas-filled detector contains a volume of compressed gas (normally a noble gas such as helium or argon) between two electrodes. A potential difference (voltage) is applied between the electrodes. When ionising radiation interacts with the atoms of the gas, it will form ion pairs due to the ionisation process. The positive ions (cations) are attracted to the negative electrode, whereas the negative ions (anions) are attracted to the positive electrode. In most detectors, the cathode is the wall of the container, and the anode is an insulated wire placed at the centre of the container. When the electrons reach the anode, they will travel through the circuit to the cathode, where they recombine with the cations. The electrical current that is formed by the flow of the electrons can then be measured with a sensitive ammeter. The response is proportional to the ionisation rate. Hence, the activity, exposure or dose rate can be calculated.

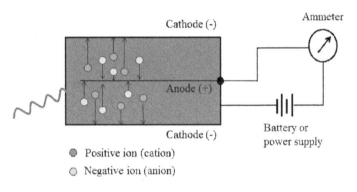

FIGURE 3.3   Basic principle of a gas-filled detector.

Note: This type of ionisation chamber is used in the radioactivity calibrator in radiopharmacy units and nuclear medicine departments. The activity can be displayed directly in units of radioactivity (Bq or Ci) after accounting for the nature of the decay emissions and sample geometry.

3.5 NOBLE GAS

PROBLEM

Explain why noble gas is commonly used in gas-filled detectors.

Solution

Noble gases, such as argon and xenon, are commonly used in gas-filled detectors because of their inert nature. Chemical reactions will not occur within a noble gas following ionisation events. Such reactions could change the characteristics and performance of the detector.

Note: The noble gas is normally placed under pressure within the chamber. Loss of pressure can affect the performance of the detector.

3.6 GAS-FILLED DETECTORS

PROBLEM

Figure 3.4 shows the logarithmic plot of the number of electrical charges collected after a single interaction as a function of the applied voltage.

    a. Name and describe the mechanisms that happen in the regions A, B, C, D and E.

    b. Give estimated applied voltage ranges for A through E.

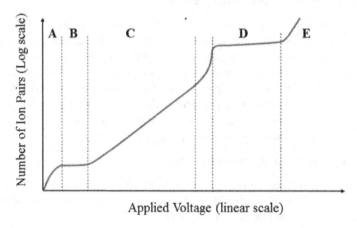

FIGURE 3.4   Plot of number of ion pairs formed versus applied voltage.

Solution

    a. **Region A: Recombination region**

        The applied voltage is relatively low in this region. When a small voltage is applied, ion pairs form where electrons are attracted to the anode, and positive ions are attracted to the cathode before they are recombined. As the voltage increases, more ions are collected and fewer ions are recombined. The current increases as the voltage increases until a saturation point.

Region B: Ionisation chamber region

As the voltage increases, a plateau is reached. In this region, the applied voltage is sufficiently high to collect almost all ion pairs. An additional increment in the applied voltage does not significantly increase the current. This is the region where the ionisation chamber operates.

Region C: Proportional region

Beyond the ionisation region, the collected current increases again as the applied voltage increases. In this region, the electrons approaching the anode are accelerated to such a very high kinetic energy that they can cause additional ionisation. This mechanism is known as gas multiplication. The amount of amplification increases as the applied voltage increases. This is the region where the proportional counter operates.

Region D: Geiger-Müller region

Beyond the proportional region, the amount of charges collected from each event is the same regardless of the amount of energy deposited by the interaction. The applied voltage is very high that even a minimally ionising particle will produce a very large current. The initial ionisation produced by the radiation triggers a complete gas breakdown as an avalanche of electrons accelerates towards and spreads along the anode. This region is called the Geiger-Müller region and is the underlying principle of the Geiger counter.

Region E: Spontaneous discharge

At this region, the applied voltage is too high since this causes continuous discharge of the gas; hence, a gas-filled detector cannot be operated at this region.

b. Estimated applied voltage range for A through E:

A: $0 - 200$ V
B: $200 - 400$ V
C: $400 - 800$ V
D: $1000 - 1400$ V
E: above 1400 V

Note: In nuclear medicine departments, ionisation chambers are used to assay the activity of radiopharmaceuticals, and Geiger-Müller detectors are used for contamination monitoring, particularly of beta emitters.

3.7 RADIONUCLIDE ACTIVITY (DOSE) CALIBRATOR

PROBLEM

a. What is an activity calibrator? Does the activity calibrator measure the patient dose contributed by a radionuclide?

b. Sketch the internal structure of an activity calibrator and explain its principle of operation.

Solution

a. The activity calibrator is a well-typed ionisation chamber used to measure the activity of administered radiopharmaceuticals. The activity calibrator does not measure the radiation dose contributed by the radionuclide, but it measures the amount of radioactivity contained within the 'dose' of a radiopharmaceutical drawn up to be administered to a patient.

Note: An activity calibrator is commonly known as the 'dose' calibrator. Strictly speaking, the term 'dose' should not be used since dose is measured in Gy; however, it is a common term still used by many people.

b. Figure 3.5 shows the internal structure of an activity calibrator. An activity calibrator comprises an ionisation chamber, high-voltage power supply, electrometer, a radionuclide selector, and a display unit. The ionisation chamber contains pressurised gas (such as argon gas), and the hermetically sealed chamber contains two co-axial cylindrical electrodes, which are connected to a high-voltage supply. The wall of the chamber is connected to the cathode, and the collector electrodes are connected to the anode. When a sample is placed into the chamber, the gas is ionised. The ion pairs then migrate towards the anode and cathode and generate an electrical current. The current is proportional to the activity of the sample. An electrometer is used to measure the very small current (about a few μA). A calibration factor is applied according to the selected radionuclide to convert the current to activity. The output is usually displayed in MBq or mCi. The activity calibrator is operated in current mode; therefore, dead-time effect is minimal. It can accurately assay activity up to 74 GBq (or 2 Ci); however, it is relatively insensitive and cannot measure activity less than 3.7×10^5 GBq (or 1 μCi).

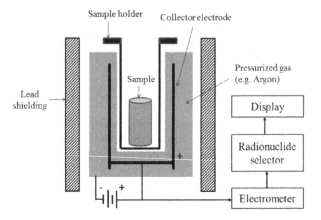

FIGURE 3.5 Sketch of a radionuclide activity calibrator.

3.8 FACTORS AFFECTING MEASUREMENT ACCURACY OF AN ACTIVITY CALIBRATOR

PROBLEM

What are the factors that may affect the measurement accuracy of an activity calibrator?

Solution

Factors that may affect the measurement accuracy of an activity calibrator include:

- Energy and abundance of the photons produced by the radionuclide.
- Geometry of the detector, for example cylindrical versus rectangular shape.
- Geometry of the radioactive source, including the vial or container that is used, such as syringe or vial size, type of material (plastic or glass) and the thickness of the container's wall.
- Condition of the activity calibrator which can be evaluated through periodical quality control tests.
- Loss of pressure of the gas contained within the chamber.

- Inadequate shielding from other radioactive source and background radioactivity.
- Ambient pressure and temperature that should not be changed abruptly.

3.9 QUALITY CONTROL OF AN ACTIVITY CALIBRATOR

PROBLEM

List the common quality control tests that should be carried out on an activity calibrator.

Solution

The common quality control tests include:

- Precision and accuracy tests using long half-life standard radioactive sources, for example, Cs-137.
- Linearity of the activity response over the range of activities used.
- Reproducibility test (or constancy test) carried out over time.
- Background activity and shielding check.
- Geometry dependence test (only during installation).

Note: The calibrator is usually checked with a long-lived standard source (e.g. Cs-137) on a daily basis to ensure correct operation when measuring the activity administered to patients.

3.10 BASIC PRINCIPLES OF SCINTILLATION DETECTORS

PROBLEM

Explain briefly the basic operating principles of a scintillation detector.

Solution

Scintillators are materials that emit visible or ultraviolet light after absorbing ionising radiation. When ionising radiation interacts with a scintillator, the electrons of the scintillator atoms are raised to an excited energy level. Eventually, these electrons will fall back to its original energy state, with the emission of visible or ultraviolet light. A scintillation detector is obtained when a scintillator is coupled to an electronic light sensor, such as a photomultiplier tube (PMT), photodiode or silicon photomultiplier. The PMT absorbs the light emitted by the scintillator and reemits it in the form of electrons via a photoelectric effect. The subsequent multiplication of these electrons results in an electrical pulse, which can then be analysed and correlate to the intensity of the radiation that struck the scintillator. In all scintillators, the amount of light emitted increases in direct proportion to the energy deposited in the scintillator.

Note: Most scintillators have more than one mode of light emission, and each mode has its characteristic decay constant. For example, luminescence is the emission of light after excitation, fluorescence is the prompt emission of light, and phosphorescence (or commonly known as afterglow) is the delayed emission of light and varies from a few nanoseconds to hours depending on the material.

3.11 SCINTILLATION DETECTORS

PROBLEM

Name three scintillation detectors that are used in nuclear medicine applications.

Solution

- Gamma well counter (sample counter) for measuring activity within standard preparations and biological samples, such as blood and urine.
- Probe detector, such as a thyroid probe detector, for measuring organ uptake.
- Gamma camera (or scintillation camera) for diagnostic imaging.

3.12 BASIC PRINCIPLES OF A GAMMA WELL COUNTER

PROBLEM

Draw a schematic diagram of a gamma well counter and explain its principles of operation.

Solution

Figure 3.6 shows the schematic diagram of a gamma well counter. A gamma well counter consists of a well-shaped scintillation detector (usually NaI[TI] crystal) with a hole in the centre. A sample is placed inside the hole for increased geometric and counting efficiency. This design gives extremely high efficiency and allows detection of very low activity, that is, less than 37 Bq (or 1 nCi). The scintillation detector is coupled to a photomultiplier tube (PMT) and connected to a pre-amplifier and amplifier. When ionising radiation interacts with the scintillation crystal, lights are emitted and converted to electrons by a thin photocathode layer in front of the PMT. The electrons undergo a multiplication process in the PMT and are then sent to a pre-amplifier and amplifier to amplify the electronic signals. The signals are then sent to a pulse height analyser to discriminate the energy information. Only the signals in the selected energy window are used for counting.

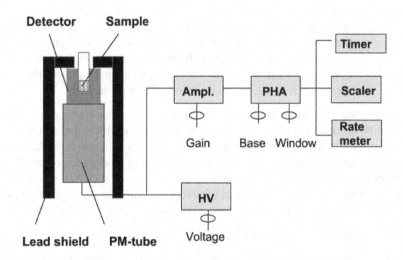

FIGURE 3.6   Schematic diagram of a gamma well counter.

3.13 CLINICAL APPLICATIONS OF GAMMA WELL COUNTER

PROBLEM

Describe briefly three clinical applications of a gamma well counter.

Solution

i. Glomerular function rate (GFR) test

The test uses a radiopharmaceutical that is rapidly excreted by the kidneys, such as Cr-51-EDTA or Tc-99m-DTPA. The total renal function is measured from the clearance of the activity in a series of blood samples taken following intravenous injection of the radiopharmaceutical. The activity of the blood samples is counted using a gamma well counter. This is a valuable test in patients undergoing chemotherapy with cytotoxic drugs that may result in renal toxicity.

ii. Plasma and red blood cells volume determination

Plasma and red blood cells volume can be determined using isotope dilution method. By diluting a known activity and volume of radiotracer in an unknown volume (plasma or red blood cells), and then measuring the activity in a volume withdrawn after adequate mixing, the unknown volume can be calculated. For plasma volume determination, I-125-Human Serum Albumin is used; whereas for red blood cell volume determination, Cr-51 is commonly used. The most common clinical application of this test is when evaluating patients for polycythemia vera.

Note: Several precautions need to be taken during the test. For example, samples should be taken with the patient in the same position each time, the radiotracer must be stably bound and the volume must be small enough as to not affect the volume being measured. In addition, equilibrium must be reached before samples are drawn; usually a 20-minute sample is used. However, it should be noted that the rate of mixing varies from patient to patient, especially in conditions such as shock or elevated haematocrits in which mixing may be delayed considerably.

iii. Radioimmunoassay (RIA)

RIA is an *in-vitro* assay technique used to measure the concentrations of an antigen (e.g. hormone levels in the blood) with very high sensitivity. In this technique, a known quantity of an antigen is labelled to a radioisotope, for example I-125. The radiolabelled antigen is then mixed with a known amount of antibody for that antigen. As a result, the radiolabelled antigen and antibody will be bound together. A sample of serum from the patient containing an unknown quantity of the same antigen is then added. This causes the unlabelled (or 'cold') antigen from the serum to compete with the radiolabelled (or 'hot') antigen. When the concentration of 'cold' antigen increases, the amount of 'cold' antigen bound to the antibody will also increase. This reduces the ratio of antibody-bound radiolabelled antigen to free radiolabelled antigen. The bound radiolabelled antigens are then separated from the unbound ones, and the radioactivity of the free radiolabelled antigen remaining in the supernatant is measured using a gamma well counter. A binding curve is then generated, which allows the amount of antigen in the patient's serum to be derived.

Note: RIA is an old assay technique, but it is still widely used due to its simplicity and sensitivity.

3.14 BASIC PRINCIPLES OF A THYROID PROBE

PROBLEM

Draw a schematic diagram of a thyroid probe and explain its principles of operation.

Solution

Figure 3.7 shows the schematic diagram of a thyroid probe. A thyroid probe consists of a cylindrical NaI(Tl) crystal coupled to a photomultiplier tube (PMT) which in turn is connected to a pre-amplifier and other associated electronics, such as amplifier, pulse height analyser (PHA), timer, scaler, rate meter and a recorder. The operation principle is similar to a gamma well counter except that a long cylindrical or conical collimator is applied on the thyroid probe to limit the field-of-view (FOV) of the thyroid. This reduces the background radiation from areas outside the thyroid to reach the detector. The probe is connected to a high-voltage power supply.

When ionising radiation interacts with the NaI(Tl) crystal, electrons are raised to an excited energy state. Eventually the electrons will return to their ground energy state by emitting the excessive energy in terms of visible or ultraviolet light. The light photons will interact with the photocathode located at the entrance of the PMT and are converted to electrons. The electrons are then accelerated and focused towards a series of dynodes in the PMT. The increasing voltage electrons will then impinge the anode. The resultant output signal will then be amplified and analysed by the PHA. Only the energies in the selected energy range will be used for counting. The timer, scaler and rate meter are used for counting of the radioactivity signals.

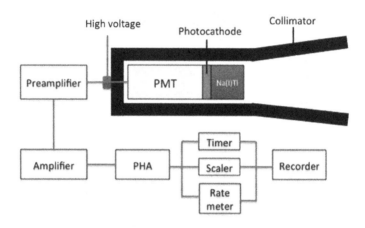

FIGURE 3.7   Schematic diagram of a thyroid probe.

3.15 THYROID UPTAKE MEASUREMENT

PROBLEM

Describe briefly the methods used for thyroid uptake measurement including 'two-capsule' and 'one-capsule' methods.

Solution

Thyroid uptake measurement can be performed using one or two capsules of I-123 or I-131 sodium iodide. A neck phantom, consisting of a Lucite or Perspex cylinder of diameter similar to the neck and containing a hole parallel to its axis for radioiodine capsule placement is required.

Two-capsule method:

Two radioiodine capsules with almost identical activities are used. Firstly, each capsule is placed in the neck phantom and counted. Secondly, one capsule is given to the patient, and another one is used as the 'standard'. The gamma emissions from the patient's neck are counted at 4–6 hours after administration and repeated at 24 hours post-administration. During each measurement, the patient's distal thigh and background activities are also obtained with the same duration of counting. The measurements are repeated with the 'standard' capsule placed in the neck phantom for counting. Finally, the uptake is calculated for each neck measurement as follows:

$$\text{Uptake} = \frac{(\text{Thyroid count} - \text{Thigh count})}{(\text{Count of standard in phantom} - \text{Background count})}$$

$$\times \frac{\text{Initial count of standard in phantom}}{\text{Initial count of patient capsule in phantom}}$$

One-capsule method:

A single radioiodine capsule is used. The capsule is first counted in the neck phantom, then ingested by the patient. As in the two-capsule method, the patient's neck, distal thigh and background activities are counted for the same period of time at 4–6 hours and again at 24 hours post-administration. The times of the capsule administration and neck counts are recorded. Finally, the uptake is calculated as follows:

$$\text{Uptake} = \frac{(\text{Thyroid count} - \text{Thigh count})}{(\text{Count of capsule in phantom} - \text{Background count})} \times e^{-0.693t/t_{1/2}}$$

where:
   $t_{1/2}$ = physical half-life of the radionuclide
   t = time elapsed between the measurement of capsule in the phantom and the patient's neck

Note: The single-capsule method reduces the cost of the examination and requires fewer measurements, but it is more susceptible to instability of the equipment, technologist errors and dead-time effects.

Instrumentation for Gamma Imaging

4

4.1 X-RAY VERSUS GAMMA-RAY IMAGING

PROBLEM

Compare and contrast the principles of image formation between projection X-ray and gamma scintigraphic imaging.

Solution

Figure 4.1 illustrates the differences between projection X-ray and gamma scintigraphic imaging. In X-ray imaging, the radiation beams are emitted from a single source (X-ray tube) in a fan or cone shape. The X-ray beams are projected in one direction only, that is, from the X-ray tube to the detector. Each radiation beam will be attenuated by the tissues as it passes through the body and reaches the detector. The final image formed on the detector is therefore representing the attenuation properties of the tissues/organs.

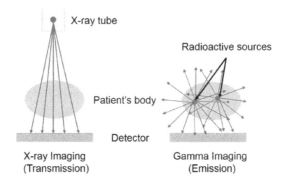

FIGURE 4.1   Comparison between X-ray imaging and gamma scintigraphic imaging.

In gamma scintigraphic imaging, radiopharmaceutical is administered to the patient's body via injection, ingestion or inhalation. The radiopharmaceutical will be distributed in the body with uptake by the targeting tissues/organs. Therefore, scintigraphic imaging constitutes multiple sources of radiation at various locations in the patient's body. The radiation beams are emitted isotopically from the patient's body and reach the detector. A collimator is necessary to allow only parallel beams to pass through the collimator and reach the detector to form a clinically useful image.

X-ray imaging provides mostly anatomical information based on tissue attenuation; on the other hand, gamma scintigraphic imaging is organ specified and is based on the tissue uptake of the radiopharmaceutical.

4.2 GAMMA CAMERA SYSTEM

PROBLEM

Figure 4.2 shows the schematic diagram of a typical gamma camera system.

a. Name the components labelled with **A**, **B** and **C** in the diagram.
b. Explain the mechanism that takes place in **C** to form a clinically useful gamma camera image.

Solution

    a. A: Scintillating detector or sodium iodide doped with thallium
       (NaI[Tl]) detector.
       B: X-, Y-positioning circuit.
       C: Energy discriminator circuit or pulse height analyser (PHA).
    b. Energy discriminator circuit (known as pulse height analyser) is
       used to filter the scattered photons that contain energies outside of
       the pre-selected energy window(s). Firstly, the four output signals
       ($\pm X$ and $\pm Y$) from the X-, Y-positioning circuit are weighted and
       summed to form a voltage pulse (known as the 'Z' pulse), which
       represents the intensity of a scintillation. Secondly, the Z-pulse is
       then sent to the energy discriminator circuit and analysed if it falls
       within the pre-set energy window(s) (with upper and lower energy
       levels). In theory, scattered or coincidence photons can be rejected
       this way because they fall outside the energy window. However, the
       filtration is not perfect since the energy loss due to Compton scat-
       tering is not very large, and the energy window cannot be too strict
       as other factors, such as time and sensitivity need to be taken into
       account.

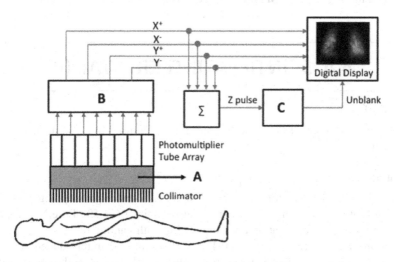

FIGURE 4.2 Schematic diagram of a typical gamma camera system.

Note: The majority of clinical gamma cameras operate using scintillation crystals and photomultiplier tubes. New solid-state detectors such as cadmium zinc telluride (CZT) are now entering into clinical use.

4.3 GAMMA RAYS DETECTION

PROBLEM

Describe briefly the mechanisms that take place when 140 keV gamma rays strike a thallium-activated sodium iodide (NaI[TI]) crystal.

Solution

When 140 keV gamma rays strike on a NaI(Tl) crystal, the gamma rays will interact with the crystal via coherent scattering, photoelectric absorption or Compton scattering, whereby the NaI molecules are raised to higher energy states through ionisation and excitation. The ionised or excited atoms then return to the ground state by emitting light photons. The light output is proportional to the energy of the incident radiation. Approximately 38 light photons are produced per 1 keV of deposited energy. The light photons then enter the photomultiplier tubes (PMTs) which are attached to the NaI(Tl) crystal.

4.4 SCINTILLATING CRYSTAL (I)

PROBLEM

Describe the physical properties of NaI(Tl) crystal used in a gamma camera system.

Solution

NaI(Tl) has relatively high density (3.67 g·cm^{-3}) and high stopping power coefficient due to the high atomic number of iodine (Z = 53). Pure NaI does not produce any scintillation after interacting with gamma ray at room temperature. A trace amount (0.1%–0.4%) of thallium is therefore added as an activator. The activating material also influences the wavelength (colour) of the light produced by the scintillator. In NaI(Tl), the scintillation light is blue (λ_{max} = 415 nm). NaI(Tl) crystals are hygroscopic and fragile. Once the crystal absorbs water, it would cause a colour change that reduces light transmission to the photomultiplier tubes (PMTs). Hence, the crystal is sealed in an aluminium container in a gamma camera system with a Pyrex glass backing to allow the light to be transmitted to the PMTs.

4.5 SCINTILLATING CRYSTAL (II)

PROBLEM

What is the difference between a scintillation-based gamma camera and a solid-state gamma camera?

Solution

In a scintillation-based gamma camera, incident gamma rays deposit their energy in the scintillator where it is converted into visible (or near-UV) light photons, which are detected by the photomultiplier tubes. In a solid-state gamma camera (e.g. cadmium zinc telluride [CZT] detector), radiation deposits its energy in the crystal lattice where it results in the generation of pairs of charge carriers (direct conversion). The application of an electric field causes the charge carriers to be swept to the cathode and anode of the device where they induce a current pulse that can be detected and positional information collected. The collimators in these systems are registered to the detector array to maximise efficiency.

4.6 FUNCTION OF THE COLLIMATOR

PROBLEM

Why is a collimator necessary in a gamma camera system?

Solution

The gamma emissions from the patient's body are isotropic; hence, a collimator is necessary to form a useful image by permitting gamma rays approaching the camera from certain directions to reach the detector while absorbing most of the scattered photons. Without a collimator, the acquisition of all the emitted radiation would result in an almost incoherent flood image.

Note: Collimators consist of a series of holes in a lead plate that allows the gamma rays to pass through to the crystal. In this way, it behaves a little like a lens on a camera.

4.7 COLLIMATOR DESIGN

PROBLEM

What are the main factors to be considered in the design of a gamma camera collimator?

Solution

- Sensitivity—Longer, narrower holes reduce the photon transmission, hence reduce sensitivity of the system.
- Resolution—Longer, narrower holes reduce the angle of acceptance, hence increase resolution of the image.
- Energy of the radioisotope—A higher energy source needs thicker septa to block the scattered radiation again resulting in loss of sensitivity.

Note: Collimators are designed for specific gamma energies. Usually below 180 keV are considered low energies (e.g. Tc-99m), between 180 and 370 keV are considered medium energies (e.g. In-111), and 370–550 keV are considered high energies (e.g. I-131). In each case, a high-resolution or high-sensitivity collimator can be used depending on the clinical application required.

4.8 TYPES OF COLLIMATORS

PROBLEM

With the aid of diagrams, describe briefly the design and application of four common types of gamma camera collimators.

Solution

- Parallel-hole collimator: It consists of a lead plate with all of the holes parallel to each other. The function is to allow only photons travelling in a direction perpendicular to the detector to reach the crystal. The image:object dimension ratio is 1:1. It is the most commonly used collimator in gamma imaging (see Figure 4.3).

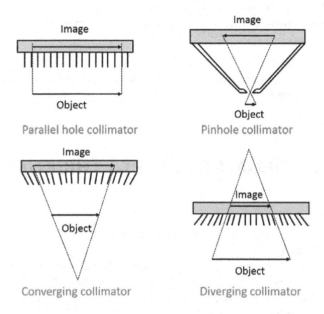

FIGURE 4.3 Four common types of gamma camera collimators.

- Converging collimator: It is a flat, multi-hole collimator in which all holes converge to a focal point at about 40–50 cm from the collimator. The focal point is normally located in the centre of the field-of-view (FOV). The image appears larger (magnified) compared to the object. The magnification increases as the object is moved away from the collimator. It is usually used to image small organs such as adrenal glands.
- Diverging collimator: When the converging collimator is flipped over, it becomes a diverging collimator whereby the focal point is now behind the collimator. It produces a minified image, and the amount of minification increases as the object is moved away from the collimator. It is usually used to image a large portion of a patient on a small FOV camera (e.g. a mobile camera).
- Pinhole collimator: It consists of a small hole (typically 3–5 mm diameter) in a piece of lead or tungsten mounted at the apex of a leaded cone. It produces a reversed and magnified image. The magnification decreases as the object is moved away from the pinhole.

When the distance between the object and the pinhole is the same as the distance between the pinhole and the crystal, there is no magnification. If the object is moved yet farther from the pinhole, the object will be minimised. The pinhole collimator is most commonly used in thyroid imaging and paediatric nuclear medicine.

Note: On modern gamma cameras, most magnification can be carried out using software zoom applications. As a result, most collimators used are of a parallel-hole design.

4.9 COLLIMATOR SEPTA THICKNESS

PROBLEM

a. How is the thickness of the septa determined in a gamma camera collimator?
b. Calculate the septa thickness of a lead collimator required for high-energy (364 keV) detection. The diameter of the collimator hole is 0.25 cm, and the septa length is 2.5 cm, given that the linear attenuation coefficient of lead at 364 keV is 2.49 cm^{-1}.

Solution

a. The collimator of a gamma camera is typically designed for less than 5% photon transmission. The thickness of the septa is determined according to the purpose of the study; for example thicker or longer septa are used for higher spatial resolution image. However, a minimum thickness of the septa is required to form a useful image which can be calculated using the following equation:

$$t \geq \frac{\dfrac{6d}{\mu}}{L - \left(\dfrac{3}{\mu}\right)}$$

where:
t = thickness of the septa
d = diameter of the hole
L = length of the septa

μ = linear attenuation coefficient of the absorber, usually lead. μ depends on the atomic number (Z) and density (ρ) of the material, as well as photon energy.

b. $t \geq \dfrac{\dfrac{6d}{\mu}}{L - \left(\dfrac{3}{\mu}\right)}$

$$t \geq \frac{\dfrac{6(0.25\,\text{cm})}{2.49\,\text{cm}^{-1}}}{2.5\,\text{cm} - \left(\dfrac{3}{2.49\,\text{cm}^{-1}}\right)}$$

$$t \geq \frac{0.602\,\text{cm}^2}{1.295\,\text{cm}}$$

$$t \geq 0.465\,\text{cm}$$

Therefore, the thickness of the collimator septa should be at least 0.465 cm.

4.10 MINIFICATION FACTOR FOR DIVERGING COLLIMATOR

PROBLEM

What is the minification factor for a diverging collimator 5 cm thick, with the distance between the front of the collimator to the convergence point of 50 cm, and a source located 10 cm away from the collimator? If the object size is 30 cm, what is the image size at this distance?

Solution

$$\text{Minification factor is expressed as } \frac{I}{O} = \frac{(f - t)}{(f + b)}$$

where:

I and O = image and object size
f = distance from the front of the collimator to the divergence point
t = thickness of the collimator
b = front of the collimator to the source

Therefore,

$$\frac{I}{O} = \frac{(50\,\text{cm} - 5\,\text{cm})}{(50\,\text{cm} + 10\,\text{cm})} = 0.75$$

$$\frac{I}{30\,\text{cm}} = 0.75$$

$$I = 22.5\,\text{cm}$$

Therefore, the image size is 22.5 cm.

Note: A diverging collimator contains holes that diverge from the detector face. The holes diverge from a point (divergence point) typically 40–50 cm behind the collimator, projecting a minified, non-inverted image of the source onto the detector.

4.11 MAGNIFICATION FACTOR FOR CONVERGING COLLIMATOR

PROBLEM

What is the magnification factor for a converging collimator 5 cm thick, with the distance between the front of the collimator to the convergence point of 40 cm, and a source located 15 cm away from the collimator? If the object size is 10 cm, what is the image size at this distance?

Solution

$$\text{Magnification factor,}\quad \frac{I}{O} = \frac{(f+t)}{(f+t-b)}$$

where:
  I and O = image and object size
  f = distance from the front of the collimator to the convergence point
  t = thickness of the collimator
  b = front of the collimator to the source

Therefore,

$$\frac{I}{O} = \frac{(40\,\text{cm} + 5\,\text{cm})}{(40\,\text{cm} + 5\,\text{cm} - 15\,\text{cm})} = 1.5$$

$$\frac{I}{10\,\text{cm}} = 1.5$$

Therefore, I = 15 cm

Note: A converging collimator contains holes that converge to a point at about 40–50 cm in front of the collimator. A converging collimator projects a magnified, non-inverted image of the source onto the detector.

4.12 PHOTOMULTIPLIER TUBE (PMT) ,

PROBLEM

Sketch the inner structure of a photomultiplier tube and explain the process of multiplication in the tube.

Solution

- A photomultiplier tube (PMT) consists of an evacuated glass tube containing a photocathode, typically 10–12 electrodes (known as dynodes) and an anode (see Figure 4.4).
- The PMT is fixed on to the NaI(TI) crystal with the photocathode facing the crystal.
- The photocathode is usually an alloy of cesium and antimony that releases electrons after absorption of light photons.
- A high-voltage power supply of approximately 1000 V is applied between the photocathode and anode, with a series of resistors dividing the voltage into equal increments.
- When a light photon strikes the photocathode, photoelectrons are emitted via the photoemission process. The electrons are attracted to the first dynode and are accelerated to the kinetic energies equal to the potential difference between the photocathode and the first dynode. For example, if the potential difference is 100 V, the kinetic energy of each electron is 100 eV.
- Approximately five electrons are ejected for each electron that strikes on the dynode. These electrons are then attracted to the next dynode, reaching kinetic energy equal to the potential difference between the first and second dynodes, and causing five electrons to be ejected for each electron that strikes on it.

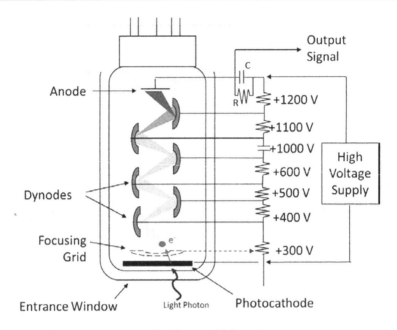

FIGURE 4.4   Inner structure of a photomultiplier tube.

- This multiplication process continues until the last dynode. The total amplification of the PMT is the product of the individual multiplication at each dynode.

Note: New smaller PMTs are now available in compact camera systems. In some cases, position sensitive PMTs (PSPMTs) are used.

4.13 X-, Y-POSITIONING CIRCUIT

PROBLEM

Describe the function(s) of the X-, Y-positioning circuit in a gamma camera system.

Solution

Scintillation events produced in the NaI(Tl) crystal are detected by a large number of photomultiplier tubes (PMTs) which are arranged in a 2D array. There are typically 30–90 PMTs in a modern gamma camera. The output voltages generated by the PMTs are fed to a position circuit which produces four output signals, namely X^+, X^-, Y^+ and Y^-. These position signals contain information about where the scintillations were produced within the crystal, as well as the intensity of each scintillation. The intensity information can be derived from the position signals by summing up the four position signals (X^+, X^-, Y^+ and Y^-) to generate a voltage pulse (known as the Z-pulse) which represents the intensity of a scintillation. The Z-pulse is then sent to the pulse height analyser (PHA) to filter the scattered and coincident photons.

4.14 ENERGY DISCRIMINATION CIRCUIT

PROBLEM

Explain the function of an energy discrimination circuit in a gamma camera system.

Solution

Although a collimator can block >99% of the scattered radiation, there may be conditions that some of the scattered photons escape the septa and reach the detector. These 'false' signals are illustrated in Figure 4.5.

The gamma photons that reach the detector after passing through the collimator may undergo three possible interactions: (1) some may be absorbed in the NaI(Tl) crystal, (2) some scatter from the crystal and (3) some pass through the crystal without any interaction. The energy discrimination circuit filters the scattered or coincidence photons by summing the signals from all the photomultiplier tubes (PMTs) (Z-pulse) and check if it falls within the pre-set energy window. In principle, the Compton scattered photons will

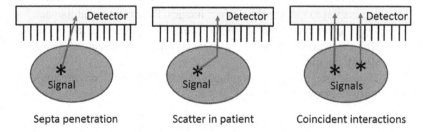

Septa penetration   Scatter in patient   Coincident interactions

FIGURE 4.5 Illustration of 'false signals' that reach the NaI(Tl) detector.

lose energy, and the coincidence interactions will contribute extra energy. A pulse height analyser (PHA) is used to filter the signals that fall outside of the preselected energy window or 'channels'; hence, only the signals that fall within the selected energy will be registered as a 'true' and count in the image.

4.15 DIGITAL DATA ACQUISITION

PROBLEM

Digital data in nuclear medicine are acquired in either frame or list mode.

    a. Describe briefly the frame and list modes acquisition techniques.
    b. Discuss the advantages and disadvantages of frame and list modes.

Solution

    a. In frame mode, a matrix with size approximate to the area of the detector is defined. A position (X, Y) in the detector corresponds to a pixel position in the matrix. Before acquisition, all pixels within the matrix are set to zero. During acquisition, every time when a new signal is detected, it is added to the corresponding (X, Y) pixel in the matrix. The data acquisition continues until a preselected time or total count is reached. Therefore, in frame mode, the size and depth of the matrix, number of frames per study and time per frame or total counts must be specified.

       In list mode, the (X, Y) signals are stored in a list instead of being immediately formed into an image. Each signal is coded with 'time mark' according to the time it is detected and stored as individual event. After acquisition is completed, the data can be sorted and displayed to form the desired images. Data can be manipulated by changing matrix size, time per frame, physiological markers, etc. after the acquisition. The 'bad' signals can also be discarded manually by the user.

    b. The advantages and disadvantages of frame and list modes are summarised below:

	ADVANTAGES	DISADVANTAGES
Frame Mode	The data are immediately displayed in a designated image format; hence, limited post-processing work is required. It requires lesser memory and disk space for acquisition and storage compared to the list mode.	The matrix (image) format must be specified before acquisition, and no manipulation of the data is possible after the image is formed.
List Mode	It provides flexibility to manipulating data according to the way the users want it to be displayed.	It generates a large amount of data which requires larger memory for acquisition, greater processing power and disk space for storage.

4.16 DIGITAL IMAGES IN NUCLEAR MEDICINE

PROBLEM

a. How is the pixel size in a gamma camera digital image determined?

b. An image generated by a gamma camera has a dimension of 40×40 cm^2. If a matrix size of 128×128 is used during image acquisition, what is the size of each pixel?

Solution

a. A profile is obtained from two small radioactive sources placed at a known separation, or from a single source moved to known positions on the X or Y axis of the camera. The profile will give two spatial peaks. The distance between the maxima of the two peaks is measured in terms of pixels and used to calculate the distance between the two source positions. Knowing the distance and the number of pixels allows calculation of the individual pixel size.

b. $\text{Pixel size} = \dfrac{40 \text{ cm}}{128 \text{ pixel}} = 0.3 \text{ cm per pixel}$

4.17 EFFECTS OF MATRIX SIZE AND STATISTICAL NOISE

PROBLEM

500,000 counts are collected in a matrix size of 128 × 128 and 256 × 256, respectively. What will be the difference in term of image noise? How many counts do we need to attain in the matrix size of 256 × 256 in order to achieve the same noise level in the 128 × 128 image?

Solution

For 128 × 128 matrix size:

$$\frac{500,000 \text{ counts}}{128 \times 128 \text{ pixels}} = 30.5 \text{ counts per pixel}$$

$$\text{Statisical noise} = \frac{100\%}{\sqrt{30.5}} = 18\%$$

For 256 × 256 matrix size:

$$\frac{500,000 \text{ counts}}{256 \times 256 \text{ pixels}} = 7.6 \text{ counts per pixel}$$

$$\text{Statisical noise} = \frac{100\%}{\sqrt{7.6}} = 36\%$$

Hence, the matrix size of 256 × 256 has 50% more noise than the matrix size of 128 × 128.

To achieve the same noise level in 256 × 256 and 128 × 128 matrix sizes:

$$\frac{\text{Total counts}}{256 \times 256 \text{ pixels}} = 30.5 \text{ counts per pixel}$$

Therefore, total counts = 1,998,848

4.18 STATIC STUDY

PROBLEM

Describe briefly the acquisition method used in static planar imaging. Give three clinical imaging examples of static study.

Solution

In static study, a single image is acquired for either a pre-set time interval or until the total numbers of counts reaches a pre-set number. The data is acquired in frame mode, and matrix size is predefined before acquisition. Static planar imaging is used for studies in which the distribution of the radiopharmaceutical is effectively static throughout the acquisition. It provides information, such as organ morphology (size, shape and position), and regions of increased or decreased uptake.

Clinical imaging examples of static imaging studies include Tc-99m-HDP bone scan, I-123-iodide thyroid scintigraphy and renal imaging using Tc-99m-DMSA.

4.19 DYNAMIC STUDY

PROBLEM

Describe briefly the acquisition method used in a dynamic study. Give three clinical imaging examples of dynamic studies.

Solution

In a dynamic study, a series of images are acquired in sequence one after another, for a pre-set time per image. This is usually done when the biokinetics of the tracer is relatively rapid. The frame rate (frame per second) and matrix size can be varied between images; however, the patient's position cannot be changed during the whole course of study.

The clinical imaging examples of dynamic studies include oesophageal transit, hepatobiliary imaging and renogram.

4.20 GATED STUDY

PROBLEM

Describe briefly the acquisition method used in a gated study. Give three clinical imaging examples of a gated study.

Solution

A gated study is performed when the dynamic process of an organ occurs too rapidly to be effectively captured by a dynamic image acquisition, and when the dynamic process is repetitive, such as the cardiac or respiratory cycle. Gated acquisitions require a physiological monitor that provides a trigger pulse to mark the beginning of each cycle of the process being studied. For example, in a gated cardiac study, an electrocardiogram (ECG) monitor provides a trigger pulse to the computer whenever the monitor detects a QRS complex. Using this technique, the image data from each desired phase of the cycle can be stored in specific location (known as 'bins'). When all the data in a bin are added together, the image represents the specific phase of the cycle.

Examples of gated studies include cardiac blood pool study for calculation of left ventricular ejection fraction, gated SPECT myocardial perfusion study for regional wall uptake and motion and a respiratory gated study of the lungs.

SPECT and PET Imaging

5

5.1 PHYSICAL PRINCIPLES OF SPECT

PROBLEM

Explain briefly the basic principles of single-photon emission computed tomography (SPECT).

Solution

SPECT is performed using a gamma camera that consists of one or more detectors. Each detector is comprised of a collimator, scintillation crystal, array of photomultiplier tubes (PMTs) and digital positional network. During SPECT imaging, the detectors are rotated around the patient at small-angle increments, typically 3°–10°. The rotation can be in continuous motion or using the 'step-and-shoot' method.

In most cases, a full 360° rotation is used to obtain an optimal reconstruction; however, in cardiac imaging, a rotation of 180° is sufficient. Individual planar images are acquired at each angle for a set period of time, for example 30 seconds. In each detector, gamma rays pass through the collimator and are absorbed by the crystal. The gamma ray energies are converted into light photons and detected by the PMTs. The images are displayed in the form of a digital matrix, for example, a 128×128 or 256×256 matrix. Each pixel contains position and count rate information. The two-dimensional (2D) planar images are pre-filtered and reconstructed using either an iterative process or filtered back projection to form a three-dimensional (3D) image. The data may then be manipulated to show thin slices along any chosen axis (e.g. transverse, sagittal or coronal) of the body. The count rates of each voxel are commonly displayed using a colour scale.

5.2 COMPARISON OF SPECT AND PLANAR IMAGING

PROBLEM

Discuss the advantages and disadvantages of SPECT in comparison to the conventional planar scintigraphic imaging.

Solution

In planar imaging, the radioactivity in the tissues in front and behind of an organ reduces contrast. If the activity in these overlapping structures is not uniform, the pattern of activity distribution from different structures will be superimposed. This causes a source of structural noise that hinders the ability to distinguish the activity distribution in the tissues. SPECT imaging can greatly improve the subject contrast and reduce structural noise by eliminating the activity in overlapping structures. In addition, SPECT imaging also allows partial correction of tissue attenuation and photon scattering in the patient.

The possible disadvantage of SPECT includes the slight decrease of spatial resolution compared to planar imaging. This is because the detector is usually closer to the patient in planar imaging than in SPECT; however, the increase in image information at depth easily compensates for any loss of resolution. In addition, using a shorter time per angle in SPECT may necessitate the use of a lower resolution/higher sensitivity collimator to obtain an adequate number of counts within the time frame.

5.3 SPECT DATA ACQUISITION

PROBLEM

Describe briefly how gamma ray energy affects the following features in SPECT imaging:

    a. The choice of a camera collimator.
    b. The use of attenuation correction in SPECT reconstructed images.

Solution

    a. A collimator is required to project the gamma photons onto the detector crystal. Collimators are essentially constructed as lead plates with holes or lead foils to channel the photons. Any photon not running parallel to the hole will be absorbed by the septa. For diagnostic imaging, collimators are classified according to three energy ranges:

ENERGY LEVEL	ENERGY RANGE (KeV)	EXAMPLES OF RADIOISOTOPES
Low energy	100–200	Tc-99m (140 keV) I-123 (159 keV)
Medium energy	200–300	In-111 (171 and 245 keV)
High energy	300–400	I-131 (364 keV)

The higher the energy, the thicker the lead septa are required to attenuate the photon. Hence, the overall sensitivity is reduced. In addition, the spatial resolution is also reduced at higher energy.

b. For SPECT imaging, the camera rotates around the patient. The peripheral margins of the patient are closer to the detector, and hence the photons originating from that area will be subjected to less attenuation than those arising from the centre of the patient. By defining the size of the patient from the outline of the image (an X-ray computed tomography [CT] image may be used), attenuation correction could be applied to correct for photon attenuation with depth.

Gamma ray attenuation is expressed as:

$$N = N_o e^{-\mu t}$$

where:

N = number of transmitted photons
N_o = number of incident photons
μ = linear attenuation coefficient of the tissue
t = thickness of the tissue

The attenuation effect will be more pronounced for lower energy gamma rays, that is, Tc-99m (140 keV). The resulting image data will therefore have a lower count density in the centre of the patient in comparison to the periphery. This can be corrected by modifying the image data using the appropriate attenuation coefficient for the energy of the gamma rays used (i.e. Tc-99m = 0.12). In this way, counts are added to the central regions to give a corrected uniform image.

5.4 PRINCIPLE OF NOISE FILTERING USING THE FOURIER METHOD

PROBLEM

Explain the principle of noise filtering using the Fourier method in SPECT image reconstruction.

Solution

The Fourier noise-filtering method is done in the spatial domain. It is used to eliminate the blurring function (star artefact) from simple back projection. In this method, the amplitude of different frequencies is modulated. The principle is to preserve the broad structures (the image) represented by low frequency and remove the fine structure (noise) represented by high frequency. There are a number of filters classified according to their functions. For example, a 'low pass filter' (e.g. Butterworth, Parzen, and Weiner filters) rejects the high-frequency data, and a 'high pass filter' (e.g. ramp filter) rejects the low to medium frequency data. Often the low pass filters are applied as pre-filters to remove high-frequency noise, and the ramp filter is then applied during back projection to remove star artefact. This combination of filter is called 'high pass-low pass filter', for example, Ramp-Parzen filter. All of the Fourier filters are characterised by a maximum frequency, known as the Nyquist frequency. The Nyquist frequency is calculated as 0.5 cycle per pixel, which is the highest fundamental frequency useful for imaging. The cut-off frequency is the maximum frequency the filter will pass. If the cut-off frequency is higher than the Nyquist frequency, the data will be filtered at the Nyquist frequency.

5.5 PRINCIPLE OF NOISE FILTERING USING THE CONVOLUTION METHOD

PROBLEM

Explain briefly the principle of the convolution method that is used in filtered back projection reconstruction.

Solution

The convolution method is carried out in the spatial domain. It uses a smoothing 'kernel', which is basically a mathematical function to remove the 1/r function, essentially by taking some counts from the neighbouring pixels and putting them back into the central pixel of interest. The '1/r' function is the geometrically blurring effect as a function of distance away from the point source, where r is the radial (distance from a point source). The most commonly used kernel in filtered back projection is the 'nine-point smoothing' kernel. An example of a smoothing calculation is illustrated below:

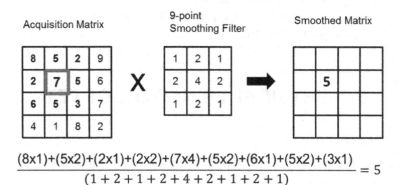

$$\frac{(8\times1)+(5\times2)+(2\times1)+(2\times2)+(7\times4)+(5\times2)+(6\times1)+(5\times2)+(3\times1)}{(1+2+1+2+4+2+1+2+1)} = 5$$

In the nine-point smoothing method, the individual pixel value in the acquisition matrix is multiplied with a 3×3 smoothing matrix. The sum of the product is then divided by the sum of all pixel values of the smoothing matrix. In this example, the pixel value '7' is converted to '5' after smoothing. Similarly, all pixel values in the acquisition matrix are smoothed, and a smoothed image is produced.

5.6 IMAGE PROCESSING USING ITERATIVE RECONSTRUCTION

PROBLEM

Explain briefly the iterative reconstruction method used in image processing of tomographic images.

Solution

- In the iterative method, an initial estimate activity is made of each element in the N × N matrix, and values are compared to the measured values.
- Corrections are applied according to the estimated values.

- If an estimated value is smaller than the measured value, each pixel that contributes to this pixel will be raised in counts, and a comparison is made again until the estimated value is achieved.
- The calculation of projection images applies the point spread function of the scintillation camera, which takes into account the decreasing spatial resolution with distance from the camera face.
- The point spread function can be modified to incorporate the effects of photon scattering in the patient. This technique is known as expectation maximisation.

5.7 PHYSICAL PRINCIPLES OF PET

PROBLEM

Describe the physical principle of positron emission tomography (PET).

Solution

Figure 5.1 illustrates the basic principle of positron emission tomography. In PET imaging, a positron emitting radiopharmaceutical, such as F-18-FDG, is administered into the patient's body. The positrons lose most of their energy in matter by causing ionisation and excitation. When a positron has lost most

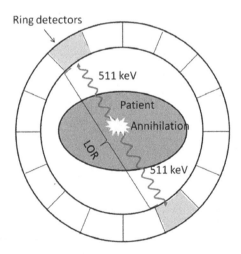

FIGURE 5.1 Basic principle of positron emission tomography (LOR = Line-of-response).

of its energy, it interacts with an electron by annihilation, whereby the entire mass of the electron-positron pair is converted into two 511 keV gamma photons, which are emitted in nearly opposite directions. The photons are then detected by a ring of detectors surrounding the patient. The ring detectors use the same fundamental detector system as the one used in the gamma camera, which include scintillating crystals and photomultiplier tubes. The detector electronics are linked so that two photons unambiguously detected within a certain time window (typically 6–12 ns) may be registered as a coincident event and originate from the same annihilation. Since the annihilation photons are emitted at almost 180° to each other, it is possible to locate their source along a straight line of coincidence (also known as the line-of-response [LOR]) by calculating the distance between each photon with the source according to their arrival time at each detector. This principle is known as the 'time-of-flight' (TOF). The coincidence events are then being stored in arrays corresponding to projection through the patient and reconstructed using standard tomographic techniques, such as filtered back projection and iterative reconstruction.

Note: A CT scanner is usually attached to a PET scanner for tissue attenuation correction, image fusion and diagnostic purposes.

5.8 ANNIHILATION COINCIDENCE DETECTION

PROBLEM

With the aid of a diagram, discuss briefly the annihilation coincidence detection (ACD) method used in PET imaging.

Solution

Figure 5.2 illustrates the principles of coincident annihilation detection method used in PET imaging. When a positron is emitted by nuclear transformation, it transverses through matter and loses energy. After it loses most of its energy, it interacts with an electron, resulting in two 511 keV photons that are emitted in nearly opposite directions. This phenomenon is known as annihilation. When the two photons are simultaneously detected within a ring of detectors surrounding the patient, it is assumed that the annihilation event occurred on the line connecting the interactions (i.e. the line-of-response [LOR]).

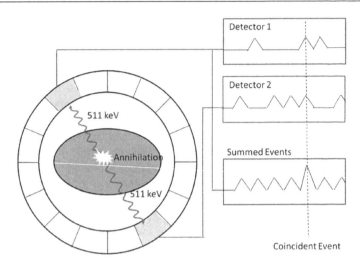

FIGURE 5.2   Coincident annihilation detection.

Each detector will generate a timed pulse when it detects an incident photon. These pulses are then combined in coincidence circuitry, and if the pulses fall within a short time window (within approximately 5ns), they are deemed to be coincident and the signal will be registered in the image.

5.9 TRUE, SCATTER AND RANDOM COINCIDENCE EVENTS

PROBLEM

Explain with the aid of diagrams, the true, scatter and random coincidence events that are possibly detected in a PET scanner.

Solution

    a. True coincidence: the two 511 keV photons that are detected from opposite detectors are from the actual source location (Figure 5.3).

    b. Scatter coincidence: the photons are scattered in the patient's body before being detected by the PET detectors. The annihilation coincidence detection (ACD) method does not present the actual location of the source.

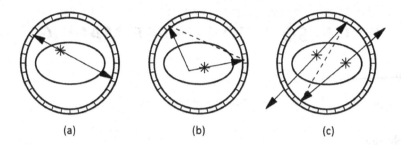

(a)   (b)   (c)

FIGURE 5.3   Illustrations of a) true, b) scatter, and c) random coincidence.

c. Random coincidence: the detectors detect several photons from different sources at the same time (or within 12 ns) and wrongly pair them as a true coincidence event, and therefore does not present the actual location of the source.

5.10 TIME-OF-FLIGHT IN PET IMAGING

PROBLEM

Explain how a PET scanner localizes the source of the coincident photons using the method called 'time-of-flight'. What are the disadvantages of using this method in PET imaging?

Solution

A PET scanner measures time-of-flight under the assumption that the location of the annihilation can be localised along the line of flight of the coincident photons by measuring time of arrival of each photon at the opposing crystal. Unless the event occurs in the exact centre of the ring, one of the photons will arrive before the other. Typical coincidence time is 12 ns. The time difference is proportional to the difference in distance and can be used to calculate the position of the event.

The disadvantages of this method include lower accuracy in determining the location of the source due to possibilities of scatter events and random coincidence, as well as reducing spatial resolution of the image.

5.11 RESOLUTION OF PET IMAGING

PROBLEM

What are the main factors affecting the resolution of PET images?

Solution

The resolution of the PET image is determined mainly by the detector size. Smaller detectors give better resolution. The ring diameter and the detector material also affect the intrinsic resolution of the PET image. The energy of the positron plays a minor role as higher energy positrons travel farther away from the site of origin before annihilating.

5.12 2D VERSUS 3D PET IMAGING

PROBLEM

What is the difference between 2D and 3D acquisition of PET data, and what are their relative advantages and disadvantages?

Solution

In 2D acquisition, coincidence events are collected by detectors in a single ring around the source, and each detector is in coincidence with any other detector in the ring and defines a 2D plane. In 3D acquisition, coincidence events are collected by detectors in multiple rings or in 2D arrays of detectors around the source. Each detector is in coincidence with any other detectors in all the rings or in the entire opposing 2D array and helps to define a 3D image. 2D acquisition is less prone to scatter coincidences and requires less electronics for volumetric acquisition. 3D acquisition is more prone to false coincidence effects and requires fast processing electronics.

5.13 COMPARISON OF Tc-99m AND F-18

PROBLEM

Compare the physical properties of Tc-99m and F-18, which are used in SPECT and PET imaging, respectively.

Solution

Comparison of physical properties of Tc-99m and F-18:

	Tc-99m	*F-18*
Production method	Mo-99 generator	Cyclotron
Mode of decay	Isomeric transition (88%)	Positron emission (97%)
	Internal conversion (12%)	Electron capture (3%)
Decay product	Tc-99	O-18 (stable)
Physical half-life	6.02 hours	110 min
Detection photon energy	140 keV	511 keV
Detection principle	Single photon detection	Dual photon (511 keV each) travelling in opposite directions

5.14 COMPARISON OF IMAGING TECHNIQUES BETWEEN SPECT AND PET

PROBLEM

Compare the SPECT and PET scanners in terms of the following:

a. Principle of projection data collection
b. Transverse image reconstruction

    c. Radionuclide
    d. Spatial resolution
    e. Attenuation

Solution

	SPECT	PET
a. Principle of projection data collection	Using physical collimator	Annihilation coincidence detection
b. Transverse image reconstruction	Filtered back projection or iterative reconstruction	Filtered back projection or iterative reconstruction
c. Radionuclide	Any radionuclides emitting X-rays, gamma rays or annihilation photons.	Positron emitters only
d. Spatial resolution	Depends on collimator and camera orbit. Within a transaxial image, the resolution in the radial direction is relatively uniform, but the tangential resolution is degraded towards the centre. Typically, 10 mm full width at half maximum (FWHM) at the centre for a 30-cm diameter orbit using Tc-99m. Larger camera orbits produce lower spatial resolution.	Relatively constant across transaxial image, best at the centre. Typically 4.5–5 mm full width at half maximum (FWHM) at the centre.
e. Attenuation	Attenuation is less significant. Attenuation correction methods are available.	Attenuation is more significant. Accurate attenuation correction is possible with a transmission source.

5.15 COMPARISON OF SPATIAL RESOLUTION AND DETECTION EFFICIENCY BETWEEN SPECT AND PET

PROBLEM

Contrast the spatial resolution and detection efficiency between SPECT and PET scanners based on the collimation and annihilation coincidence detection (ACD) methods.

Solution

In SPECT, the spatial resolution and the detection efficiency are primarily determined by the collimator. Both are ultimately limited by the compromise between collimator efficiency and collimator spatial resolution that is a consequence of collimated image formation. Larger camera orbits result in lower resolution. This causes the spatial resolution to deteriorate from the edge to the centre in transverse SPECT images.

In comparison, PET has better spatial resolution and detection efficiency. ACD method is used as an electronic collimator for the PET scanner by determining the path of the detected photons. ACD allows detection of two annihilation photons given off by a single positron annihilation event. The two photons (511 keV) are considered to be from the same event if they are counted within a defined coincidence timing window and recorded as occurring along the line-of-response (LOR). The location of the event can then be calculated using the principle of time-of-flight (TOF). This combination approach effectively filters the scattered photons and determines the location of the source, hence improving the spatial resolution and detection efficiency of PET imaging.

Imaging Techniques in Nuclear Medicine

6

6.1 WHOLE BODY BONE SPECT IMAGING

PROBLEM

Explain the basic principle, radiopharmaceutical used, image acquisition, image processing and data interpretation in a whole-body bone single-photon emission computed tomography (SPECT) imaging.

Solution

Principle

- Whole body bone SPECT imaging is a sensitive method for detecting a variety of anatomical and physiological abnormalities of the musculoskeletal system. A Tc-99m-labelled radiopharmaceutical (such as methylene diphosphonate [MDP]), which localises in the bone, is injected intravenously to the patient. The activity accumulated in the bone is then evaluated through the SPECT images. It helps to diagnose a number of bone conditions, including bone cancer or metastasis, inflammation, fractures and bone infection.

Radiopharmaceutical

- The most commonly used radiopharmaceutical for bone SPECT imaging is Tc-99m labelled with methylene diphosphonate (MDP).

MDP absorbs onto the hydroxyapatite mineral of the bone; hence, it can be used as a marker of bone turnover and bone perfusion.

- Tc-99m-MDP is injected intravenously to the patient with typical administered activity of 555–800 MBq for adults. Administered activity for children is determined based on body weight. Generally, the administered activity for children and adolescents is 9.3 MBq per kg with a minimum administered activity of 37 MBq.

Image Acquisition

- SPECT imaging is performed at two to three hours following injection of the radiopharmaceutical to allow optimal concentration of the radiotracer uptake by active osteoblasts. The delay also allows time for background activity to be excreted from the body. Patients are required to empty their bladders before the scan.
- A low-energy high-resolution (LEHR) collimator is used. The acquisition energy window is set to $140\% \pm 10\%$ keV and a matrix size of 256×1024 is typically used. The patient is positioned in the supine, arms by the side and feet first position. Anterior and posterior views are acquired simultaneously if a dual-head camera is used. The total imaging time is approximately 30 minutes. Patients are asked to stay still during the scanning.

Image processing and data interpretation

- The tomographic images are reconstructed using either filtered back projection or iterative method.
- In whole-body bone SPECT, about half of the radiopharmaceutical is localised in the bones. The more active the bone turnover, the more radionuclide concentration (hot spot) will be seen. However, abnormal radionuclide concentration should not be confidently assigned to a particular pathology without a typical pattern of radiotracer distribution, such as multiple randomly placed foci in metastatic bone disease. In the absence of this, a correlation of radiotracer uptake with alternative imaging modalities such as radiography, computed tomography (CT), or magnetic resonance imaging (MRI), should be reviewed when available, as this can significantly increase the accuracy of bone scintigraphy interpretation.

Note: SPECT-CT hybrid imaging has been demonstrated to significantly improve the accuracy of bone scintigraphy interpretation when the CT component may be justified for indeterminate lesions and complex joint abnormalities, such as the spine and feet.

6.2 CARDIAC IMAGING

PROBLEM

A cardiologist would like to determine the left ventricular ejection fraction of a patient's heart at the end diastole and end systole cycles.

    a. Suggest the best radionuclide imaging technique for this patient.
    b. With the aid of a diagram, explain the principle and techniques of the examination you suggested above.

Solution

Multiple-gated acquisition (MUGA) or gated cardiac blood pool imaging using Tc-99m-labelled red blood cells (Figure 6.1).

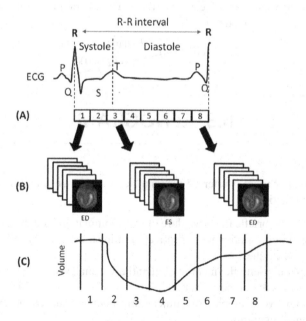

FIGURE 6.1 Basic principles of multiple gated acquisitions (MUGA) for cardiac ejection fraction study. The numbers 1 to 8 represent the frames allocated over the cardiac cycle/heart beat.

Abbreviations: ED = end diastole; ES = end systole.

- Tc-99m-labelled red blood cells are administered to the patient through intravenous injection.
- Patient is connected to an electrocardiogram (ECG) machine and imaged using a gamma camera. As the gamma camera images are acquired, the patient's heartbeat is used to 'gate' the acquisition.
- In MUGA, data are acquired in synchronisation with the R-wave.
- R–R interval on ECG, representing one cardiac cycle, is typically divided into eight frames of equal duration (A).
- Image data from each frame are acquired over multiple cardiac cycles and stored separately in specific locations ('bin') of computer memory (B).
- When all data in a bin are added together, the image represents a specific phase of the cardiac cycle.
- Typically, a volume curve is obtained, which represents the endocardial volume for each of the eight frames (C).
- The ejection fraction is the proportion of blood that is pumped from the left ventricle with each heartbeat. The ejection fraction is calculated using Equation 6.1:

$$\text{Ejection fraction} (\%) = \frac{ED(\text{net}) - ES(\text{net})}{ED \times 100} \tag{6.1}$$

6.3 RENOGRAM

PROBLEM

An adult male patient is referred for renal scintigraphy or renogram to assess his kidney function.

a. Describe briefly the procedure of this examination including patient preparation, radiopharmaceutical, administered activity and imaging technique.
b. Explain briefly the method of quantitative analysis of the renogram to assess the kidney function.
c. Name two examples of clinical information that can be derived from a renogram.

Solution

a. A renogram refers to serial imaging after intravenous administration of Tc-99m-DTPA or Tc-99m-MAG3. It is used for qualitative and quantitative evaluation of differential renal functions. The typical administered activity for an adult is 300 MBq for both kidneys and 150 MBq if there is only one kidney. The patient will be asked to drink 300–500 mL of water approximately 10–15 minutes prior to the scan. For the investigation of suspected obstructed kidneys, a diuretic furosemide (Lasix®) will be injected to promote kidney drainage. The patient should void his bladder just prior to the exam. Imaging begins immediately after injection of the radiopharmaceutical intravenously followed by a rapid saline flush. The syringe and tubing residual activity is measured and recorded for further analysis. A common technique involves dynamic acquisition of 1- to 2-second images for 1 minute (vascular phase), followed by 15- to 60-second images for 20 to 30 minutes (functional uptake, cortical transit and excretion phases).

b. The regions of interest (ROIs) are drawn around the heart and the whole kidneys but occasionally are limited to the renal cortex if a considerable amount of collecting system activity is present. A background ROI is placed adjacent to each kidney. Using the software, time-activity curves for all the ROIs are obtained. Depending on the ROIs drawn, the time-activity curves will reflect the functional clearance of the whole kidney, renal cortex or collecting system. The differential renal function is calculated based on the relative counts accumulated in each kidney during the first two minutes of the renogram. Figure 6.2 shows an example of the time-activity curve.

c. Urinary tract obstruction and renovascular hypertension.

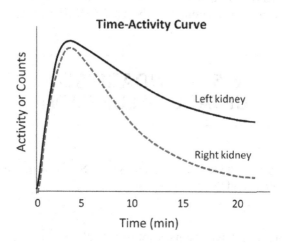

FIGURE 6.2 Example of time-activity curve that can be obtained from a renogram.

6.4 RADIOIMMUNOASSAY (RIA)

PROBLEM

Describe the principle of radioimmunoassay.

Solution

Radioimmunoassay (RIA) is a sensitive *in vitro* assay technique used to measure the concentration of an antigen (e.g. hormone level in the blood) using a specific antibody. During the test, a known amount of antigen is labelled with a radioactive source (e.g. I-125), and the radiolabelled antigen is then mixed with a known amount of antibody. The radiolabelled antigen will bind together with the antibody. Subsequently, a sample of serum from the patient containing an unknown amount of the same antigen is added. This causes a competitive binding of two antigens with the antibody. The antigen, which has a higher concentration, will bind extensively with the antibody, displacing the other.

In this case, if the antigen in the patient's serum has a higher concentration, it would bind with the antibody and displace the radiolabelled antigen. The bound antigens are then separated from the unbound ones, and the radioactivity of the unbound antigen remaining in the supernatant is measured using a gamma counter. Using known standards and formulas, the amount of antigen in the patient's serum can be derived.

Note: A number of non-radioactive methods can now be used to replace RIA, such as the enzyme-linked immunosorbent assay (ELISA).

6.5 STANDARDISED UPTAKE VALUE (SUV)

PROBLEM

Describe the use of standardised uptake value (SUV) in positron emission tomography (PET) and its limitations.

Solution

The SUV is used as a relative measure of radiopharmaceutical uptake, such as FDG, standardised for the injected activity and patient weight. An ROI is defined on the area of the PET image, and the SUV within the ROI is calculated using the following formula:

$$SUV = \frac{\text{Tracer activity in the tumour per unit mass } (Bq \cdot kg^{-1})}{\text{Amount of injected radioactivity per unit body mass } (Bq \cdot kg^{-1})}$$

The main purposes of SUVs are to provide a good correlation between the histological grades of the tumour and to facilitate comparisons between patients. However, there are several limitations of the use of an SUV. First, it has a large degree of variability due to physical and biological sources of errors. Furthermore, the suggestion of a threshold value of an SUV to characterise benign or malignancy of a lesion remains a clinical challenge. The sensitivity and specificity of using the SUV threshold usually decreases with lesions smaller than 7 mm. Another limitation of the SUV is in identifying the lesion boundaries and determining the counts within the ROI. The SUV may be influenced by image noise, low image resolution and variable user-biased ROI selection.

6.6 PET IMAGING APPLICATIONS

PROBLEM
Discuss the clinical applications of F-18-FDG PET imaging in oncology, neurology and cardiology.

Solution

In oncology, F-18-FDG PET imaging is very useful for the following applications:

- To assess regional tumour metabolic activity by evaluating the distribution and uptake as standard uptake values (tumour staging).
- To monitor the success of therapy by comparing the pre- and post-therapy images.
- To detect early recurrent tumours.
- To provide a whole-body survey for cancer that may have metastasised, as the FDG radiopharmaceutical is distributed throughout the body and whole-body imaging is commonly done without adding additional activity or dose to the patient.
- To identify benign and malignant growths.
- To guide radiotherapy dose planning.

In neurology, F-18-FDG PET imaging can provide diagnostic information of:

- Alzheimer's and other dementia conditions.
- Movement disorders, for example, Parkinson's disease.
- Location of the epileptic seizures prior to surgery.

In cardiology, F-18-FDG PET imaging is commonly used to:

- Determine the viability of heart tissue following a suspected heart attack.
- Predict the success of angioplasty (balloon) or bypass surgery.
- Determine if coronary arteries are blocked.

Note: There are many other PET radiopharmaceuticals (such as F-18-labelled thymidine [FLT], rubidium-82 chloride, nitrogen-13-ammonia and gallium-68-dotatate) that can be used for oncology, neurology and cardiac imaging; however, F-18-FDG is the currently the most widely used PET radiopharmaceutical due to its clinical efficacy, cost and availability.

Radionuclide Therapy

7

7.1 SEALED AND UNSEALED SOURCE THERAPY

PROBLEM

Differentiate between sealed and unsealed source therapy. What are the main factors to be considered when selecting a suitable radionuclide for unsealed source therapy?

Solution

Unsealed source therapy relates the use of soluble forms of radioactive sources, which are administered to the patient body by either injection or ingestion. Unsealed source therapy is commonly known as radionuclide therapy or molecular targeted radiotherapy. Sealed source therapy uses radionuclides, which are sealed in a capsule, rod or metal wire during therapy. Sealed source therapy is more commonly known as 'brachytherapy'.

The main factors to be considered when choosing a suitable radionuclide for unsealed source therapy include its effective half-life, types of radiation, energy range and its biochemical characteristics. Effective half-life includes both the physical and biological half-life. A suitable range of physical half-life is between six hours to seven days, whereas the biological half-life depends on a number of factors, such as radiotracer delivery, uptake kinetics, metabolism, clearance and excretion. For therapeutic purposes, radiations with high linear energy transfer (LET), such as alpha and beta particles, are preferred because they allow very high ionisation per length of travel and reduce the need for additional radiological protection. For biochemical characteristics, a suitable radionuclide should provide selective concentrations and prolonged retention in the tumour, while maintaining minimum uptake in the normal tissue.

7.2 THERAPEUTIC PROCEDURES IN NUCLEAR MEDICINE

PROBLEM

List five therapeutic procedures that are commonly done in nuclear medicine.

Solution

    i. Hyperthyroidism treatment using I-131-sodium iodide radiopharma-ceutical.

    ii. Thyroid carcinoma treatment using high-dose I-131-sodium iodide.

    iii. Treatment of polycythemia using P-32-phosphate.

    iv. Bone metastases palliative treatment using Sm-153-EDTMP (methylenephosphonic acid) or Ra-223-dichloride.

    v. Treatment for hepatocellular carcinoma using Y-90-labelled micro-spheres or Re-188-lipiodol.

7.3 HYPERTHYROIDISM ABSORBED DOSE CALCULATION

PROBLEM

Calculate the absorbed dose to the thyroid gland of a hyperthyroidism patient following administration of 925 MBq I-131 ($t_{1/2} = 8.0$ days), assuming 85% uptake, a biological half-life of 3 days for thyroid clearance of I-131, and S-value of 2.2×10^{-3} mGy(MBq·h)$^{-1}$.

Solution

- Effective half-life, $T_e = \dfrac{T_p \times T_b}{T_p + T_b} = 2.18$ days

- Cumulated activity, $A = 1.44\ A_0 \cdot T_e \cdot f_h = 1.44 \times 925$ MBq $\times$ 2.18 days $\times$ 0.85

$$= 2468.196 \text{ MBq·days} \times 24 \text{ h·day}^{-1}$$

$$= 59236.70 \text{ MBq·h}$$

- Absorbed dose $= A \cdot S = 59236.70$ MBq·h $\times 2.2 \times 10^{-3}$ mGy(MBq·h)$^{-1}$

$$= 130.32 \text{ mGy}$$

7.4 I-131 TREATMENT GUIDELINES

PROBLEM

Discuss the general instructions or precautions relating to radiation protection that are given to patients before, during and after the I-131 treatment.

Solution

Before the Treatment

- The patient will be given a list of prohibited food, dietary information and medication that could potentially affect the thyroid uptake of the administered radiopharmaceutical.
- Female patients who are of childbearing age must be tested for pregnancy prior to administration of the I-131, unless pregnancy is clinically proven to be impossible in the patient.
- Patients who are breastfeeding should be excluded or feeding should be carried out by someone else using bottled formula milk.
- The treating nuclear medicine physician should confirm that the patient is continent of urine or an arrangement should be made to prevent any contamination caused by urinary incontinence.
- Written informed consent must be obtained from the patient before therapy.

During the Treatment

- The patient's identity must be confirmed before administration of I-131.
- If the patient is to be treated as an inpatient, nursing personnel must be instructed on radiation safety. Selected nursing personnel must be provided with appropriate personal protective equipment (e.g. lead apron and thyroid shield) and radiation monitoring device (e.g. film badge and personal pocket dosimeter).
- Radiation monitoring of the thyroid glands of the patient should be performed periodically by the medical physicists or qualified staff.
- A patient is required to remain in the hospital until the external dose rate from the patient is below the activity limit set by the local/national radiation protection authorities.
- Any significant medical conditions should be noted, and contingency plans should be declared in case of a medical emergency.

Concern about radiation exposure should not interfere with the prompt, appropriate medical treatment of the patient should an acute medical problem develop.

Post-Treatment
- Before discharge from the hospital, the patient needs to be briefed on the methods to reduce unnecessary radiation exposure to the family members and members of the public. Written instructions should be provided to the patient.
- The patient should be advised to avoid pregnancy for a period of time stated in the local radiation protection guidelines.
- Patients must be provided with a written document stating that they have been given a radionuclide treatment and details of the radionuclide since the radioactivity may be detected by monitoring devices at borders such as airports during travel.

7.5 CALCULATION OF ADMINISTERED ACTIVITY FOR I-131 TREATMENT

PROBLEM

A patient with differentiated thyroid cancer is scheduled for I-131 therapy. The weight of the thyroid is estimated to be 30 g from the ultrasonography images. According to the thyroid uptake profile, the thyroid uptake at 24 h post-administration is 67.3%, and the effective half-life of I-131 is 4.5 days. Calculate the total activity to be administered for I-131 therapy for the prescribed dose of 100 Gy to the thyroid.

Solution

Using the Marinelli-Quimby formula,

$$\text{Administered activity (MBq)} = \frac{21.4\,AW}{UT_e}$$

where:

A = prescribed dose (Gy)
W = weight of the thyroid glands (g)
U = percentage uptake of I-131 at 24 h (%)
T_e = effective half-life (d)

Therefore,

$$\text{Administered activity (MBq)} = \frac{21.4 \ (100 \ \text{Gy})(30 \ \text{g})}{(67.3\%)(4.5 \ \text{d})} = 212 \ \text{MBq}$$

Hence, the activity of I-131 to be administered to the patient is 212 MBq.

Note: The thyroid weight can be estimated based on the volume obtained from either a thyroid single-photon emission computed tomography (SPECT) imaging or ultrasonography. The density of the thyroid gland was assumed to be 1 g·cm^{-3}. The volume of each thyroid lobe is calculated using the following formula based on an ellipsoid model:

$$\text{Volume (mL)} = \frac{\pi abc}{6}$$

where:
  a = length (cm), b = width (cm) and c = thickness (cm).

The total thyroid volume is obtained by adding the volume of both lobes.

7.6 RADIOIODINE THERAPY AND PREGNANCY

PROBLEM

What is the risk of radioiodine treatment to a foetus? Should this treatment be avoided at all time for pregnant women?

Solution

Radioiodine can easily cross the maternal placenta, and foetal thyroid uptake following therapeutic doses can pose significant problems for the foetus, particularly with respect to causing permanent hypothyroidism. As a general rule, pregnant women should not be treated with any radioactive substance unless the therapy is required to save her life. Thyroid cancers are relatively unaggressive compared to most other cancers. As a result, radioiodine treatment should be delayed until after pregnancy. However, if any therapy is to be performed during pregnancy, this should be restricted to after thyroid surgery during the second or third trimester of pregnancy.

Note: According to the ICRP Publication 84, a major problem occurs when a female, who is not thought to be pregnant, is treated for thyroid carcinoma and is found to be pregnant after the administration of radioiodine. If a patient is discovered to be pregnant shortly after a therapeutic radioiodine administration, maternal hydration and frequent voiding should be encouraged to help eliminate maternal radioactivity and to reduce radioiodine residence time in the bladder. If the conceptus is more than 8 weeks' post-conception and the pregnancy is discovered within 12 hours of radioiodine administration, giving the mother 60–130 mg of stable potassium iodide will partially block the foetal thyroid and reduce thyroid dose. After 12 hours post-radioiodine administration, this intervention is not very effective.

7.7 SAFE ADMINISTRATION OF I-131

PROBLEM

Discuss the safe oral administration of I-131 in a patient.

Solution

The safe oral administration of I-131 should be practiced as following:

    i. I-131 should be administered in a controlled area (e.g. hot-lab or the treatment room).
   ii. A radioactive waste disposal bag for contaminated items, such as tissue papers or glove, should be prepared prior to administration.
  iii. The patient is asked to sit at a table covered with adsorbent pads, and the floor beneath the patient should also be covered by adsorbent pads.
   iv. If the I-131 is administered in capsules, they should be transferred to the patient's mouth by tipping from a small shielded (>1 cm Pb) container.
    v. I-131 administered in oral solution should be sucked up through a straw by the patient from a lead-shielded vial. The vial should be flushed with water several times to ensure all the prescribed activity has been consumed. At the end of the procedure, the patient should then be asked to drink several glasses of water to clean the teeth and mouth.

7.8 GUIDANCE LEVEL FOR HOSPITALISATION OF I-131 PATIENTS

PROBLEM

What is the guidance level for maximum activity on hospital discharge for patients undergoing I-131 therapy, as recommended by the International Atomic Energy Agency (IAEA)?

Solution

The guidance level for the maximum activity for I-131 therapy patients discharge from hospital is 1,100 MBq, as recommended by the IAEA in the International Basic Safety Standard 1996.

7.9 RADIOEMBOLISATION

PROBLEM

Explain the principle, techniques and advantages of radioembolisation in cancer treatment.

Solution

Radioembolisation is a cancer treatment in which radioactive particles of specific sizes (about 15–60 μm) are delivered to the tumour through intra-arterial injection. The radioactive particles will be trapped in the microvessels and reside within the tumour volume to emit radiation that kill the cancer cells. Radioembolisation is most often used to treat unresectable or metastasised liver cancer.

Radioembolisation is usually performed under the radiological guidance of X-ray fluoroscopy. The interventional radiologists will identify and insert a catheter into the artery that supplies blood to the tumour. The radioactive particles of prescribed sizes (contained in saline solution), amount and prescribed radioactivity are then injected into the artery (or arteries) of the tumour. The radioactive particles will remain in the tumour, blocking blood flow to the cancer cells and killing them by the means of the ionising radiation.

The tumour may eventually shrink over time, and the radioactive particles will gradually decay to a stable state. The particles may stay permanently in the tissues or may be excreted by the body. Radioembolisation is often used in combination with other forms of cancer treatment, such as surgery and chemotherapy, to enhance the overall therapeutic efficacy.

The advantages of radioembolisation include minimally invasive procedures, less severe side effects, reduced complications and faster patient recovery time.

7.10 RADIOIMMUNOTHERAPY (RIT)

PROBLEM

Explain briefly the basic principles and advantages of radioimmunotherapy (RIT). Give three conditions that can be treated with RIT.

Solution

RIT is a type of cancer-cell-targeted therapy that uses monoclonal antibodies (mAbs) directed against tumour-associated antigens labelled with a therapeutic radionuclide. It is a combination of radiation therapy and immunotherapy. When injected into the blood stream, the radiolabelled mAbs will travel to and bind to the cancer cells, allowing a high radiation dose to be delivered directly to the tumour. The major benefit of RIT is the ability for the antibody to specifically bind to a tumour-associated antigen, hence increasing the radiation dose delivered to the tumour cells while decreasing the radiation dose to the surrounding normal tissues.

RIT can be used to treat non-Hodgkin's B-cell lymphoma, chronic lymphocytic leukaemia and immune diseases.

Internal Radiation Dosimetry

8

8.1 INTERNAL RADIATION DOSIMETRY

PROBLEM

Explain why the estimation of absorbed dose in internal radiation dosimetry presents greater challenges than the measurement of external radiation exposure.

Solution

The estimation of absorbed dose in internal radiation dosimetry is more complicated due to the following factors:

- The distribution of radionuclide within the body and its uptake in certain critical organs are difficult to determine.
- There is inhomogeneous distribution of the radionuclide even within the critical target organ or tissues.
- The biological half-life of the radionuclide may vary between patients and may be modified by disease and the pathological conditions.

8.2 FACTORS AFFECTING ABSORBED DOSE TO AN ORGAN

PROBLEM

List the factors that are likely to influence the absorbed dose to an organ.

Solution

Factors that are likely to influence the absorbed dose to an organ include:

- Type of radionuclide.
- Activity administered.
- Activity in the source organ.
- Activity in the other organs.
- The size and shape of the organs.
- The kinetics of the radiopharmaceutical.
- The quality of the radiopharmaceutical in terms of the radiochemical purity.

8.3 SOURCE AND TARGET ORGAN

PROBLEM

Define source and target organ used in medical internal radiation dosimetry (MIRD) estimation.

Solution

Source organ(s) is/are the organ(s) of interest with significant uptake of radioisotope.

Target organ(s) is/are the organ(s) that receives radiation from the other source organs, for which the absorbed dose is to be calculated. The source and target organs may be the same when the radiation dose, due to the radioactivity in the target itself, is calculated.

8.4 MONTE CARLO MODELLING

PROBLEM

Explain briefly the basic principle of Monte Carlo modelling in radiation dosimetry.

Solution

Monte Carlo modelling is a ray-tracing method in which the fate of an individual photon or particle is determined. The method is based on randomly sampling a probability distribution for each successive interaction. For example, Monte Carlo modelling can be used to simulate the random trajectories of individual photons or particles in a matter by using the knowledge of probability distribution governing the individual interactions of photons or particles in materials.

One keeps track of the physical quantities of interest for a large number of histories to provide the required information about the average quantities. Monte Carlo modelling in radiation dosimetry requires detailed knowledge of the absorption and scattering coefficients for the specific radiation energies and for the properties of various types of tissues.

8.5 MIRD FORMULA

PROBLEM

The medical internal radiation dosimetry (MIRD) formula is given below:

$$\overline{D}_{r_k} = \sum_h \tilde{A}_h S(r_k \leftarrow r_h)$$

$$\tilde{A}_h = A_0 \cdot f_h \cdot 1.44 T_e$$

Name all the quantities used in the formulas above and state their SI units.

Solution

SYMBOLS	QUANTITIES	SI UNITS
$\bar{D}_{r_k}$	The mean dose for a target organ, r_k	Gray (Gy)
$\sum_h$	The sum of all the source organs, r_h	No unit
$\tilde{A}_h$	Cumulated activity for each source organ, r_h	Becquerel·second (Bq·s)
$S(r_k \leftarrow r_h)$	Absorbed dose in the target organ, r_k, per unit of cumulated activity in each source organ, r_h	Gy (Bq·s)$^{-1}$
A_0	Initial activity of the radionuclide	Becquerel (Bq)
f_h	Fraction of uptake in the source organ, r_h	No unit
T_e	Effective half-life of the radionuclide	Second (s)

8.6 ABSORBED FRACTION

PROBLEM

Define the absorbed fraction, ϕ, and list the factors that determine its value.

Solution

Absorbed fraction is the fraction of energy emitted by the source organ that is absorbed in the target organ.

The value is determined by:

- The size of the source organ.
- The size of the target organ.
- The relative position of these organs in the body.
- The energy of the photons.
- The attenuation properties of the tissues between the source and the target organs.

8.7 S-VALUE

PROBLEM

Define S-value used in the MIRD formula. What factors determine the S-value?

Solution

S-value is defined as the dose to the target organ per unit of cumulated activity in a specified source organ. It is determined by:

- The mass of the target organ.
- The type and amount of ionising radiation emitted per disintegration.
- The fraction of the emitted radiation energy that reaches and is absorbed by the target organ.
- Each S factor is specific to a particular source organ/target organ combination.

8.8 ABSORBED DOSE CALCULATION (I)

PROBLEM

Calculate the absorbed dose to the thyroid gland of a hyperthyroidism patient following administration of 1110 MBq I-131 ($t_{1/2}$ = 8.0 days), assuming 60% uptake, a biological half-life of four days for thyroid clearance of I-131, and S factor of 2.2×10^{-3} mGy(MBq·h)$^{-1}$.

Solution

- Effective half-life, $T_e = (T_p \times T_b) / (T_p + T_b)$

$$= 2.67 \text{ days}$$

- Cumulated activity, $\tilde{A} = 1.44 A_0 f_h T$

$$= 1.44 \times 1110 \text{ MBq} \times 2.67 \times 0.6$$

$$= 2560.64 \text{ MBq days} \times 24 \text{ h(day)}^{-1}$$

$$= 61455 \text{ MBq} \cdot \text{h}$$

- Absorbed dose $= \tilde{A} \cdot S = 61455 \text{ MBq} \cdot \text{h} \times 2.2 \times 10^{-3} \text{ mGy(MBq} \cdot \text{h)}^{-1}$

$$= 135.2 \text{ mGy}$$

8.9 ABSORBED DOSE CALCULATION (II)

PROBLEM

Estimate the absorbed dose to the liver and bone marrow from the intravenous injection of 37 MBq of Tc-99m-sulphur colloid that localises uniformly and completely in the liver. Assume instantaneous uptake of the colloid in the liver and an infinite biological half-life. The S factors for Tc-99m are given in Table 8.1.

TABLE 8.1   S-values for Tc-99m

TARGET ORGAN	S-VALUES, $Gy(MBq \cdot s)^{-1}$
Liver	3.5×10^{-9}
Ovaries	3.4×10^{-11}
Spleen	6.9×10^{-11}
Red bone marrow	1.2×10^{-10}

Source organ: Liver

Solution

$$\tilde{A} = 1.44 \, A_0 f_h T_e$$

$$= 1.44(37\,\text{MBq})(6\,\text{h})(3600\,\text{s/h})$$

$$= 1.2 \times 10^6 \,\text{MBq s}$$

$$D = \tilde{A} S$$

$$D_{Li} = (1.2 \times 10^6 \,\text{MBq s})(3.5 \times 10^{-9} \,\text{Gy/MBq s})$$

$$= 4.2 \times 10^{-3} \,\text{Gy}$$

$$D_{BM} = (1.2 \times 10^6 \,\text{MBq s})(1.2 \times 10^{-10} \,\text{Gy/MBq s})$$

$$= 1.4 \times 10^{-4} \,\text{Gy}$$

8.10 ABSORBED DOSE CALCULATION (III)

PROBLEM

Calculate the absorbed dose to the lungs from the administration of 148 MBq Tc-99m-MAA particles, assuming that 99% of the particles are trapped in the lungs. The S-value for the lungs is 1.40×10^{-11} mGy(Bq·h)$^{-1}$. Assume that the Tc-99m activity is uniformly distributed in the lungs, and 45% of the activity is cleared from the lungs with a biological half-life of 3 hours and 55% with a biological half-life of 7 hours.

Solution

Physical half-life of Tc-99m = 6 h. The effective half-life of two biological clearances:

$$T_{e1} = (3 \times 6/3 + 6) = 2 \text{ h}$$

$$T_{e2} = (7 \times 6/7 + 6) = 3.2 \text{ h}$$

$$
\begin{aligned}
\tilde{A} &= 1.44\, A_0 f_h T_e \\
&= 1.44 \times 148 \times 10^6 \times 0.99 \times (0.45 \times 2 + 0.55 \times 3.2) \\
&= 0.56 \text{ GBq·h}
\end{aligned}
$$

$$
\begin{aligned}
D &= \tilde{A} \cdot S \\
&= 0.56 \times 10^9 \times 1.40 \times 10^{-11} = 7.89 \text{ mGy}
\end{aligned}
$$

8.11 ABSORBED DOSE CALCULATION (IV)

PROBLEM

Calculate the absorbed dose to the liver from a 111 MBq injection of Tc-99m sulphur colloid ($t_{1/2} = 6$ h). Assume that:

- 60% of the activity trapped in the liver.
- 30% activity uptake by the spleen.
- 10% activity uptake by the red bone marrow.
- There is instantaneous uptake and no biological excretion of the radiopharmaceutical.

The S factors are given as the following:

$$S \text{ (Liver} \leftarrow \text{Liver)} = 4.6 \times 10^{-7} \text{ Gy} \cdot \mu\text{Ci}^{-1} \cdot \text{h}^{-1}$$

$$S \text{ (Liver} \leftarrow \text{Spleen)} = 9.2 \times 10^{-9} \text{ Gy} \cdot \mu\text{Ci}^{-1} \cdot \text{h}^{-1}$$

$$S \text{ (Liver} \leftarrow \text{Bone Marrow)} = 1.6 \times 10^{-8} \text{ Gy} \cdot \mu\text{Ci}^{-1} \cdot \text{h}^{-1}$$

Solution

Assume there is no biological excretion of the radioisotope, hence the

$$T_e = T_p = 6 \text{ h}$$

$$\begin{aligned}
\tilde{A}_{\text{Liver}} &= 1.44 \, A_0 f_h T_e \\
&= 1.44 \, (111 \text{ MBq} \times 27 \text{ }\mu\text{Ci}\cdot\text{MBq}^{-1}) \, (0.6) \, (6 \text{ h}) \\
&= 15536 \text{ }\mu\text{Ci}\cdot\text{h}
\end{aligned}$$

$$\begin{aligned}
\tilde{A}_{\text{Spleen}} &= 1.44 \, A_0 f_h T_e \\
&= 1.44 \, (111 \text{ MBq} \times 27 \text{ }\mu\text{Ci}\cdot\text{MBq}^{-1}) \, (0.3) \, (6 \text{ h}) \\
&= 7768 \text{ }\mu\text{Ci}\cdot\text{h}
\end{aligned}$$

$$\begin{aligned}
\tilde{A}_{\text{Red Marrow}} &= 1.44 \, A_0 f_h T_e \\
&= 1.44 \, (111 \text{ MBq} \times 27 \text{ }\mu\text{Ci}\cdot\text{MBq}^{-1}) \, (0.1) \, (6 \text{ h}) \\
&= 2589 \text{ }\mu\text{Ci}\cdot\text{h}
\end{aligned}$$

$$\begin{aligned}
\bar{D}(\text{Liver} \leftarrow \text{Liver}) &= \tilde{A}_{\text{Liver}} \cdot S \\
&= 15536 \text{ }\mu\text{Ci}\cdot\text{h} \, (4.6 \times 10^{-7} \text{ Gy} \cdot \mu\text{Ci}^{-1} \cdot \text{h}^{-1}) \\
&= 7.1 \text{ mGy}
\end{aligned}$$

$$\begin{aligned}
\bar{D}(\text{Liver} \leftarrow \text{Spleen}) &= \tilde{A}_{\text{Liver}} \cdot S \\
&= 7768 \text{ }\mu\text{Ci}\cdot\text{h} \, (9.2 \times 10^{-9} \text{ Gy} \cdot \mu\text{Ci}^{-1} \cdot \text{h}^{-1}) \\
&= 0.07 \text{ mGy}
\end{aligned}$$

$$\begin{aligned}
\bar{D}(\text{Liver} \leftarrow \text{RM}) &= \tilde{A}_{\text{Liver}} \cdot S \\
&= 2589 \text{ }\mu\text{Ci}\cdot\text{h} \, (1.6 \times 10^{-8} \text{ Gy} \cdot \mu\text{Ci}^{-1} \cdot \text{h}^{-1}) \\
&= 0.04 \text{ mGy}
\end{aligned}$$

Hence, the total absorbed dose to the liver is the sum of all $\bar{D}$ above

$$\bar{D}_{\text{Liver}} = (7.1 + 0.07 + 0.04) \text{ mGy} = 7.21 \text{ mGy}$$

8.12 MIRD FORMALISM ASSUMPTIONS

PROBLEM

State the key assumptions made in the MIRD formalism.

Solution

   i. The radioactivity is assumed to be uniformly distributed in each source organ and absorbing target organ.

   ii. The organ size and geometries are simplified for mathematical computational.

   iii. The tissue density and composition are assumed to be homogenous in each organ.

   iv. The phantoms for the 'reference' adult and child are only approximations of the physical dimensions of any given individual.

   v. The energy deposition is averaged over the entire mass of the target organs.

   vi. Dose contributions from Bremsstrahlung and other secondary sources are ignored.

   vii. With a few exceptions, low-energy photons and all particulate radiations are assumed to be absorbed locally (i.e. non-penetrating).

Quality Control in Nuclear Medicine

9

9.1 QUALITY CONTROL OF DOSE CALIBRATOR

PROBLEM

List four basic quality control (QC) tests of a dose calibrator and state when are they carried out.

Solution

- Constancy test—daily.
- Accuracy test—at installation, annually, and after repairs.
- Linearity test—at installation, quarterly, and after repairs.
- Geometry test—at installation and after repairs.

9.2 EXTRINSIC AND INTRINSIC MEASUREMENT

PROBLEM

Define extrinsic and intrinsic measurement of the performance of a scintillation camera.

Solution

Extrinsic measurement is the measurement of a scintillation camera performance with collimator attached; while intrinsic measurement is the measurement of a scintillation camera performance when the collimator is removed.

9.3 QUALITY CONTROL METHODS

PROBLEM

Describe the methods for measuring the following performances of a scintillation camera:

  a. Uniformity
  b. Spatial resolution
  c. Energy resolution
  d. Spatial linearity
  e. System efficiency

Solution

  a. Uniformity is a measure of a camera's response to uniform irradiation of the detector surface. The intrinsic uniformity is measured by placing a point radioactive source (typically 3.7–18.5 MBq Tc-99m) in front of an uncollimated camera. The source should be placed at a distance more than four times the largest dimension of the crystal, and at least five times away if the uniformity image is to be analysed quantitatively. Extrinsic uniformity is assessed by placing a uniform planar radioactive source in front of a collimated camera. The planar source should be large enough to cover the crystal of the camera.

The integral and differential uniformity is calculated for both useful field-of-view (UFOV) and central field-of-view (CFOV). UFOV is the area of the scintillator surface which is defined from the inner walls of the collimator. It is typically about 90% of the full FOV. Whereas CFOV is the area of the detector surface that is 75% of the UFOV. The integral uniformity is calculated using the following formula for both FOV:

$$\text{Integral uniformity (\%)} = \frac{\text{Maximum count} - \text{minimum count}}{\text{Maximum count} + \text{minimum count}} \times 100\%$$

According to the National Electrical Manufacturers Association (NEMA) standards, the integral uniformity should be $\leq 5\%$ at 5 million counts.

The differential uniformity is calculated by counting the difference in any five contiguous pixels for row and column of pixels within the UFOV and CFOV, as follows:

$$\text{Differential uniformity (\%)} = \frac{\text{Highest} - \text{lowest difference}}{\text{Highest} + \text{lowest difference}} \times 100\%$$

b. Spatial resolution is a measure of camera's ability to accurately display spatial variations in activity concentration. High spatial resolution is the ability to distinguish separate radioactive objects in close proximity. The extrinsic spatial resolution is evaluated by acquiring an image of a line (or point) source, using a computer interfaced to the collimated camera and determining the line spread function (LSF) or the point spread function (PSF).

$$\text{The system resolution, } R_s = \sqrt{R_c^2 + R_I^2}$$

where R_c is the collimator resolution defined as the full width at half maximum (FWHM) of the radiation transmitted through the collimator from a line source.

R_I is the intrinsic resolution, which is determined quantitatively by acquiring an image with a sheet of lead containing thin slits placed against the uncollimated camera using a point source. Typical values for intrinsic resolution are from 2 to 4 mm.

c. Energy resolution is a measure of the camera's ability to distinguish between interactions depositing different energies in the detector. The energy resolution is measured by exposing the camera to a radioactive point source, emitting monoenergetic photons and acquiring a spectrum of the energy (Z) pulse. The energy resolution is calculated from the FWHM of the photopeak and expressed as a percentage of the central value of the peak. The typical value for intrinsic energy resolution is 9%–11%.

d. Spatial linearity is a measure of the camera's ability to represent the geometric shapes of an object accurately. It is determined from the images of a bar phantom, line sources or other phantom by assessing the straightness of the lines in the image. Quantitatively, assessment can be done by measuring the FWHM of the LSF of a line source.

e. System efficiency or efficiency is the fraction of radiation emitted by a source that produces counts in the image. The system efficiency (E_s) is the product of three factors:

$$E_s = E_c \times E_i \times f$$

where:

E_c = fraction of photons emitted by a source that penetrate the collimator holes.

E_i = fraction of photons penetrating the collimator that interact with the detector.

f = fraction of interacting photons accepted by the energy discrimination circuits.

9.4 UNIFORMITY

PROBLEM

Discuss the common causes of non-uniformity of a gamma camera image and ways to overcome them.

Solution

The common causes of non-uniformity of a gamma camera image include:

i. There are variations in light production from gamma ray interaction in the detector, light transmission to photomultiplier tube (PMT) and in the light detection and gains of the PMT. These variations result in photopeaks of different amplitude and FWHM. Because there are many PMTs across the detector, the amplitude and FWHM of the photopeaks will vary spatially across the detector surface. These variations in PMT response contribute greatly to the non-uniformity of the images. The non-uniformity can be corrected by adjusting the look-up table provided by the manufacturer. The calibration factor is obtained by moving a collimated radioactive source across the detector face with a constant pulse height analyser (PHA) window, and variations in the photopeak values are determined as a function of their spatial positions.

ii. Non-linearity in the X-, Y-positioning of pulses along the field-of-view: The spatial non-linearity is the systematic errors in the X-, Y-positioning coordinates resulting from local count compression and expansion. When a radioactive source is moved across from the edge to the centre of a PMT, more counts are pulled towards the centre. This effect is known as pincushion distortion. The counts are lower between the edges of adjacent PMTs, causing a cold spot in the area. The non-linearity can be corrected by using dedicated software. The correction factors are generated by calculating the spatial shift of the observed position of an event from its actual position.

iii. Edge packing: Edge packing is observed around the edge of an image as a bright peripheral ring that results in non-uniformity of the image. This is because more light photons are reflected near the edge of the detector to the PMT. Normally a 5-cm-wide lead ring (masking ring) is attached around the edge of the collimator to mask such effect.

9.5 SPATIAL RESOLUTION

PROBLEM

List and explain five factors that affect the spatial resolution of a gamma camera image.

Solution

i. The collimator hole's size and length will affect the range of accepted photon angles. Thicker or longer septa will block more scattered photons, hence increasing the spatial resolution.

ii. Scintillation crystal thickness can affect the probability of photons scattering. Although thicker crystal increases the sensitivity, it also decreases the spatial resolution of the system.

iii. Gamma ray energy may affect the probability of scatter in crystal and probability of septal penetration. Higher energies increase the probability of Compton scattering and septal penetration, hence reducing the spatial resolution.

iv. The size and number of PMTs may affect positional accuracy. Better spatial resolution can be achieved by using more and smaller-size PMTs.

v. The distance of the source-to-collimator can affect the spatial resolution. More scattered radiation can pass through the collimator holes when the source is farther away from the collimator, hence reducing the spatial resolution.

9.6 SENSITIVITY

PROBLEM

a. Define the sensitivity of a gamma camera.

b. What are the factors affecting the sensitivity of a gamma camera?

Solution

a. The sensitivity of a gamma camera is defined as the number of counts per unit time detected by the camera for each unit of activity from a radioactive source (cps·MBq^{-1}).

b. The factors affecting the sensitivity of a gamma camera include the geometric efficiency of the collimator, detection efficiency of the detector, pulse height analyser (PHA) discrimination settings and the dead time of the system.

9.7 COLLIMATOR EFFICIENCY

PROBLEM

a. Define collimator efficiency.

b. Discuss the factors affecting collimator efficiency.

Solution

a. Collimator efficiency is also known as geometric efficiency. It is defined as the number of gamma ray photons passing through the collimator holes per unit activity presents in a radioactive source.

For the parallel hole collimator, the collimator efficiency, E_c, is calculated as the following:

$$E_c = K \cdot \frac{d^4}{t_e^2 (d+a)^2}$$

where:

K = constant, which is a function of the shape and arrangement of holes in the collimator. It varies between 0.24 for round holes in a hexagonal array to 0.28 for square holes in a square array.

d = hole diameter

t_e = effective length of the collimator hole

a = septal thickness

b. The collimator efficiency, E_c, for parallel-hole collimator increases with increasing diameter of the collimator holes (d) and decreases with increasing collimator thickness (t) and septa thickness (a). For parallel-hole collimators, E_c is not affected by the source-to-detector distance for an extended planar source. However, for other types of collimators, E_c varies with the types of collimators as a function of source-to-collimator distance.

9.8 COLLIMATOR RESOLUTION

PROBLEM

Figure 9.1 shows the spatial resolution expressed in FWHM versus source-to-collimator distance for three different types of collimators. Name the collimators labelled with A, B and C according to their sensitivity and resolution. Give your reasoning for the answer.

Solution

A—Low-resolution collimator

B—Medium-resolution collimator

C—High-resolution collimator

For the same source-to-collimator distance, the low-resolution collimator has a larger FWHM.

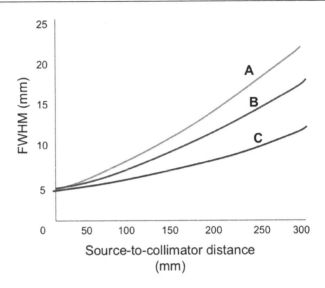

FIGURE 9.1   FWHM (mm) versus source-to-collimator distance (mm).

9.9 MODULATION TRANSFER FUNCTION (MTF)

PROBLEM

a. What is the modulation transfer function (MTF) of an imaging system?

b. If the photomultiplier tubes (PMTs) and pulse height analyser (PHA) of a gamma camera system have MTFs of 0.6 and 0.8, respectively, at a certain spatial resolution, what is the overall MTF of the system?

c. Figure 9.2 shows the plot of MTF against spatial frequency. Compare the spatial resolution of systems A, B and C.

Solution

a. MTF of an imaging system is a plot of the imaging system's modulation versus spatial frequency, where modulation is essentially the output contrast normalised by the input contrast. The principle of MTF is illustrated in Figure 9.3.

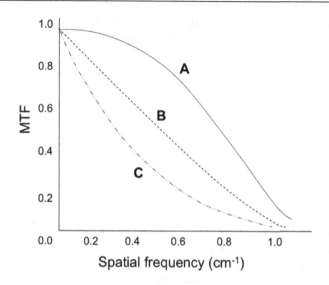

FIGURE 9.2   Plot of MTF versus spatial frequency (cm^{-1}).

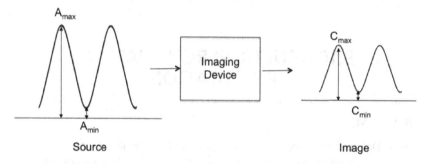

FIGURE 9.3   Principle of MTF.

Consider a source has sinusoidal distribution with peaks (maximum activity, A_{max}) and valleys (minimum activity, A_{min}), which give a spatial frequency in cycles per mm. The contrast, or modulation, in the source activity (M_s) is calculated as:

$$M_s = \frac{A_{max} - A_{min}}{A_{max} + A_{min}}$$

As no imaging device is absolutely perfect, it represents the distribution of activity in the image with C_{max} for the peak and C_{min} for the valley, which are smaller in magnitude than the A_{max} and A_{min}. Therefore, the modulation in the image (M_I) is given as:

$$M_I = \frac{C_{max} - C_{min}}{C_{max} + C_{min}}$$

The MTF at a spatial frequency, v, is then calculated as the ratio of M_I to M_S:

$$MTF(v) = \frac{M_I}{M_S}$$

b. $MTF = MTF_1 \times MTF_2 \times MTF_3 \dots$
   Therefore, the overall MTF of the system $= 0.6 \times 0.8 = 0.48$
c. System A has the highest spatial resolution compared to systems B and C, and system B has higher spatial resolution than system C.

9.10 MULTIENERGY SPATIAL REGISTRATION

PROBLEM

a. Define multienergy spatial registration of a gamma camera system.
b. How is multienergy spatial registration assessed on a gamma camera?

Solution

a. Multienergy spatial registration is a measure of the camera's ability to maintain the same image magnification, regardless of the energies deposited in the crystal by the incident photons.
b. The multienergy spatial registration can be assessed by imaging several point sources of Ga-67. Ga-67 has three useful photopeak energies for imaging at 93, 184 and 300 keV. The source is placed offset from the centre of the camera. Only one photopeak energy is tested at a time. The centroid of the count distribution of each source should be at the same position in the image for all three energies.

9.11 COLD SPOT ARTEFACT

PROBLEM

List four factors that may contribute to the formation of a cold spot artefact in a scintillation camera image.

Solution

    i. Off-centred window setting on pulse height analyser
   ii. Defective photomultiplier tube
  iii. Metal in the field-of-view (metal-like jewellery)
   iv. NaI crystal defect (such as a crack)

9.12 CENTRE OF ROTATION (COR)

PROBLEM

    a. Illustrate with the aid of a diagram, the formation of image artefact caused by misalignment of the centre of rotation (COR) in a single photon emission computed tomography (SPECT) system.
    b. What are the causes of COR misalignment?
    c. Describe the methods to assess COR of a SPECT system and ways to correct them.

Solution

    a. Ideally, the COR of a SPECT system must be aligned exactly to the centre, in the X-direction of all projection images (shown in the left image in Figure 9.4). If the COR is misaligned and not corrected, it may cause a loss of spatial resolution in the resultant transverse images. If the misalignment is large, it can cause a point source to appear like a 'doughnut' shape (as illustrated in Figure 9.4).
    b. The causes of COR misalignments include improper shifting in camera tuning (electronics), mechanics of the rotating gantry (e.g. the camera head may not be exactly centred in the gantry, refer to Figure 9.5) and misaligned attachment of the collimator to the detector.
    c. COR can be assessed by placing a point or line source within the field-of-view (FOV) of a camera with a collimator attached. The position

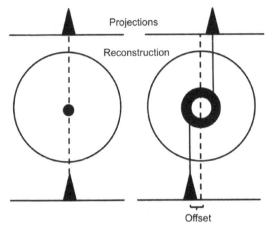

FIGURE 9.4 Formation of COR misalignment artefact.

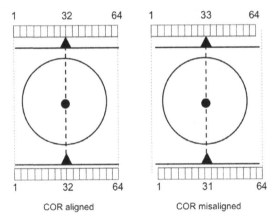

FIGURE 9.5 Misalignment of COR due to mechanics of the rotating gantry.

of the source is off-centre (at some distance away from the central axis of the detector). If a line source is used, it is placed parallel to the axis of rotation (AOR). A set of projection images are then acquired from different projection angles for 360° and stored in a 64 × 64 matrix. The position of the source along the X- and Y-axis from the centre is then computed. The X-axis plot should give a symmetrical bell-shaped curve, and the Y-axis plot should be a straight line passing through the pixel on which the source is positioned.

9.13 PARTIAL VOLUME EFFECT

PROBLEM

a. What is partial volume effect and how does this effect degrade the quality of a gamma camera image?

b. Explain briefly how partial volume effect can be corrected.

Solution

a. Partial volume effect is the loss of apparent activity in small objects or regions when the object partially occupies the sensitive volume of the imaging device (in space or time). In gamma imaging, when a hot spot relative to a 'cold' background is smaller than twice the spatial resolution of the camera, the activity around the hot object is smeared over a larger area (pixel in two dimensions [2D], and voxel in three dimensions [3D]) than it occupies in the reconstructed image. While the total counts are preserved, the object appears to be larger and has a lower activity concentration than its actual value. Analogously, a cold spot relative to a hot background would appear smaller with higher activity around the object.

b. Partial volume effect can be corrected by applying a correction factor, known as the recovery coefficient, on the image. The recovery coefficient is the ratio of the reconstructed count density to the true count density of the region of interest smaller than the spatial resolution of the system. The recovery coefficient can be determined by measuring the count densities of different objects containing the same activity but with sizes larger and smaller than the spatial resolution of the system.

9.14 SPECT QUALITY CONTROL PHANTOM

PROBLEM

Name one of the most common used multifunctional quality control (QC) phantoms in SPECT. What are the QC tests that can be performed using this phantom?

Solution

One of the most common used phantoms in SPECT QC is the 'Jaszczak' phantom. It can be used for semi-quantitative assessment of uniformity, spatial resolution and image contrast of the system. This phantom may be filled with a solution of the imaging radionuclide and is usually used for acceptance testing and periodic QC testing of SPECT systems.

Radiation Protection in Nuclear Medicine

10

10.1 RADIATION PROTECTION TERMINOLOGY

PROBLEM

Define the following terms used in radiation protection aspects:

a. Equivalent dose (H_T)
b. Effective dose (H_E)
c. Committed equivalent dose $(H_T [50])$
d. Deep dose equivalent $(H_p [10])$
e. Shallow dose equivalent $(H_p [0.07])$

Solution

a. Equivalent dose (H_T) is a measure of the radiation dose to tissue taking into consideration the different relative biological effects of different types of ionising radiation. The equivalent dose in tissue, T, is given by the expression:

$$H_T = \sum W_R \cdot D_{T,R}$$

where:

$D_{T,R}$ = absorbed dose averaged over the tissue or organ T, due to radiation R

W_R = radiation weighting factor due to radiation R

b. Effective dose (H_E) = sum of the weighted equivalent doses in all the tissues and organs of the body. It is given by the expression:

$$H_E = \sum W_T \cdot H_T$$

where:

H_T = equivalent dose in tissue or organ T

W_T = weighting factor for tissue T

c. Committed equivalent dose, H_T (50), is the time integral of the equivalent dose-rate in a specific tissue, T, following the intake of a radionuclide into the body. Unless specified otherwise, an integration time of 50 years after intake is recommended for the occupational dose and 70 years for the members of the public.

d. Deep-dose equivalent (H_p [10]) is the dose equivalent at a tissue depth of 1 cm (1000 mg·cm^{-2}) due to external whole-body exposure to ionising radiation.

e. Shallow-dose equivalent (H_p [0.07]) is the external exposure dose equivalent to the skin or an extremity at a tissue depth of 0.07 mm (7 mg·cm^{-2}) averaged over an area of 1 cm^2.

10.2 ANNUAL DOSE LIMITS

PROBLEM

State the annual dose limits (effective dose) for the following individuals as recommended by the International Commission on Radiological Protection (ICRP):

a. Radiation workers

b. Apprentices or students

c. Members of the public

Solution

a. 20 mSv per year, averaged over defined periods of five years, with no single year exceeding 50 mSv.
b. 6 mSv per year.
c. 1 mSv per year.

10.3 CLASSIFICATION OF RADIATION WORK AREAS

PROBLEM

Define the following classification of radiation work areas:

a. Clean area
b. Supervised area
c. Controlled area

Solution

a. Clean area is a work area where the annual dose received by a worker is not likely to exceed the dose limit for a member of the public, that is, 1 mSv·y^{-1}.
b. Supervised area is a work area for which the occupational exposure conditions are kept under review, even though specific protective measures and safety provision are not normally needed. The area must be demarcated with radiation warning signs, and legible notices must be clearly posted.
c. Controlled area is a work area where specific protection measures and safety provisions could be required for controlling normal exposures or preventing the spread of contamination during normal working conditions and preventing or limiting the extent of potential exposures. Annual dose received by a worker in this area is likely to exceed 3/10 of the annual occupational dose limit. The area must be demarcated with radiation warning signs, and legible notices must be clearly posted.

10.4 EXPOSURE RATE

PROBLEM

Calculate the exposure rate at 1 m from a vial containing 1110 MBq of Tl-201, given that the exposure rate constant, Γ, of Tl-201 is 0.45 R·cm^2 (mCi·h)$^{-1}$ at 1 cm.

Solution

$$\text{Exposure rate} = \frac{\Gamma A}{d^2}$$

where:
Γ = gamma constant (R·cm^2 [mCi·h]$^{-1}$)
A = activity of the source (mCi)
t = time (h)
d = distance from the source (cm)

Therefore,

$$\text{Exposure rate} = \frac{0.45\,\text{R} \cdot \dfrac{\text{cm}^2}{\text{mCi} \cdot \text{h}} \left(1110\,\text{MBq} \times \dfrac{\text{mCi}}{37\,\text{MBq}} \right)}{(100\,\text{cm})^2}$$

$$= 0.00135 \ \text{R·h}^{-1}$$
$$= 1.35 \ \text{mR·h}^{-1}$$

10.5 HALF VALUE LAYER (HVL)

PROBLEM

Calculate the number of half value layers (HVLs) and the thickness of lead (Pb) required to reducing the exposure rate from a 370 GBq Ir-192 source to less than 10 mR·h^{-1} at 1 m from the source, given that the exposure rate constant, Γ, for Pb is 2.2 R·cm^2 (mCi·h)$^{-1}$ at 1 cm and the HVL is 3 mm.

Solution

$$\text{Exposure rate at 1 m} = \frac{2.2\dfrac{R \cdot cm^2}{mCi \cdot h}\left(370 \text{ GBq} \times \dfrac{27 \text{ mCi}}{GBq}\right)}{(100 \text{ cm})^2}$$

$$= 2.198 \text{ R/h} = 2198 \text{ mR·h}^{-1}$$

$$\frac{2198}{10} \text{ mR·h}^{-1} = 219.8 \text{ mR·h}^{-1}$$

A factor of 220 is required to reduce the exposure rate from 2198 to 10 mR·h⁻¹. In terms of HVL,

$$2^n > 220$$

While $2^7 = 128$ and $2^8 = 256$, the number of HVLs required to reduce the exposure rate to 10 mR·h⁻¹ is 8.

Given that 1 HVL = 3 mm, therefore the thickness of Pb required to reduce the exposure rate to 10 mR·h⁻¹ is 8 × 3 mm = 24 mm.

10.6 ACCIDENTAL EXPOSURE TO HIGH ACTIVITY SOURCE

PROBLEM

A nurse accidentally walked into a shielded room where a 3.7 TBq Ir-192 source was left exposed for testing purposes. The nurse was unaware of the source, and she was standing approximately 1 m away from the source for about 15 minutes. What is her estimated whole-body dose, given that the gamma constant for Ir-192 is 1.3 mSv·h⁻¹ at 1 cm per MBq?

Solution

An approximate dose from a small source can be calculated as following:

$$\text{Dose} = \frac{\Gamma A t}{d^2}$$

where:

Γ = gamma constant (mSv·cm^2 [MBq·h]$^{-1}$)
A = activity of the source (MBq)
t = time (h)
d = distance from the source (cm)

Therefore,

$$\text{Dose} = \frac{1.3\frac{\text{mSv}}{\text{MBq}\cdot\text{h}}\left(3.7\times10^6\,\text{MBq}\right)\left(15\,\text{min}\times\frac{h}{60\,\text{min}}\right)}{(100\ \text{cm})^2} = 120\ \text{mSv}$$

Note: Although this is an approximation, it is adequate given the usual uncertainty with the individual's exact distance from the source and the time spent near it. While the radiation dose has exceeded the annual dose limit for a radiation worker (20 mSv per year averaged for a period of five consecutive years), this incident would result in no significant acute medical consequence to the personnel, as the acute radiation syndrome is only noticeable for a dose of approximately 2 Sv. Considering the stochastic effects, a dose of approximately 1 Sv would increase the chance of cancer by 5% compared to the non-exposed population.

10.7 FOETAL DOSE CALCULATION FOR HIGH-DOSE RADIONUCLIDE THERAPY

PROBLEM

A patient with thyroid carcinoma was administered with a diagnostic dose of 185 MBq I-131 for a whole-body scan. Thyroid uptakes measured at 2 h and 24 h are 5% and 24%, respectively. Three days later, a therapy dose of 2.035 GBq was administered. Percentage activities retained by the whole body at 1 day and 2 days are 25% and 15%, respectively. The physical half-life of I-131 is 8 days while the biological half-life is assumed to be 80 days.

a. Two weeks after the radioiodine therapy, the patient discovered that she was pregnant. How would you estimate the gestation age?
b. Assuming that the foetus was four weeks old during the administration of I-131, what is the estimated dose to the foetus? Given that the S-value (uterus ← thyroid) at 12th week is

$$3.70 \times 10^{-17}\ \text{Gy (Bq·s)}^{-1}$$

c. Based on the dose estimated in (b), do you advise termination of pregnancy or continuation to full term? Give your justification.

d. If the pregnancy was discovered within 12 hours post-administration of I-131, what would be your advice to reduce the foetal dose?

Solution

a. The gestation age can be estimated from the first day of the last menstrual period (LMP) of the mother. If the LMP is unknown or not regular, then the gestation age can be estimated using ultrasound measurements of the foetus combined with the dates of first foetal heart tones and other developmental milestones.

b. Effective half-life, $T_e = (T_p \cdot T_b)(T_p + T_b)^{-1}$

$$= (8\ d \times 80\ d)(8\ d + 80\ d)^{-1}$$

$$= 7.3\ d$$

$$= 63720\ s$$

The cumulated activities, $\tilde{A}$, for different uptake fractions are calculated as following:

AT TIME (h)	INITIAL ACTIVITY, A_0 (Bq)	FRACTION UPTAKE, f_h		T_e (s)	$\tilde{A} = A_0 \cdot f_h \cdot 1.44 T_e$ (Bq·s)
2	1.85×10^8	0.05	1.44	630,720	8.40×10^{12}
24	1.85×10^8	0.24	1.44	630,720	4.03×10^{13}
24	2.04×10^9	0.25	1.44	630,720	4.62×10^{14}
48	2.04×10^9	0.15	1.44	630,720	2.77×10^{14}
Total $\tilde{A}$					**7.88×10^{14}**

Hence, the total accumulated dose to the embryo up to 12th week is estimated to be:

$$D = \tilde{A}.S$$

$$= (7.88 \times 10^{14}\ \text{Bq·s})(3.70 \times 10^{-17}\ \text{Gy[Bq·s]}^{-1})$$

$$= 29.2 \times 10^{-3}\ \text{Gy}$$

$$\sim 30\ \text{mGy}$$

c. I would not recommend termination of pregnancy for this patient. The estimated threshold doses for the embryo and foetus as recommended by the International Atomic Energy Agency (IAEA) are shown in the table below:

AGE	THRESHOLD DOSE FOR LETHAL EFFECTS (mGy)	THRESHOLD DOSE FOR MALFORMATIONS (mGy)
1 day	100	No effect
14 days	250	—
18 days	500	250
20 days	>500	250
50 days	>1000	500
50 days to birth	>1000	>500

The foetal dose estimated in this case is approximately 30 mGy at age of ~16 days, which is lower than the threshold dose recommended by the IAEA. Lethal effects and malformations of the foetus are not expected. Furthermore, the foetal thyroid accumulates iodine only after about 10 weeks of gestational age. The International Commission on Radiological Protection (ICRP) also stated termination of pregnancy at foetal dose of less than 100 mGy is not justified based upon radiation risk.

d. If the pregnancy was discovered within 12 hours post-administration of I-131, prompt oral administration of stable potassium iodide to the mother is recommended to reduce foetal thyroid dose. This may need to be repeated several times with the advice from the nuclear medicine consultant.

Note: It is imperative to rule out pregnancy prior to the administration of radioiodine therapy due to the potential detrimental side effects of foetal exposure. The dose of I-131 in this particular case would be unlikely to result in any deterministic effects. An accurate pregnancy screening protocol may be warranted in preventing inadvertent I-131 treatment in early pregnancies. In many countries, it is common practice to perform a pregnancy test prior to high-dose I-131 scanning or therapy for women of childbearing age unless there is a clear history of prior tubal ligation or hysterectomy precluding pregnancy. It is unlikely that the inadvertent administration of a diagnostic radiopharmaceutical to a pregnant patient would result in any action other than patient counselling. As a result of good practice, it is extremely rare to have to take any remedial action following the administration of therapeutic radiopharmaceuticals.

10.8 RADIATION WORKERS DURING PREGNANCY

PROBLEM

A nuclear medicine worker is concerned about radiation exposure during her pregnancy. What would be your advice to the worker as well as to the department?

Solution

For most nuclear medicine diagnostic procedures, there is no necessity for pregnant staff to take any additional precautions other than limiting their direct contact to the radioactive source. As the radiation exposure from the patients who have been administered with radiopharmaceuticals is quite low, there is no radiological reason for the pregnant worker to refrain from undertaking the imaging procedures. According to the Basic Safety Standards, notification of pregnancy shall not be considered a reason to exclude a female worker from work. However, the pregnant worker should refrain from handling the high-dose therapeutic radioactivities, such as I-131 therapy for thyroid cancer.

10.9 RADIATION WORKERS DOSE LIMIT

PROBLEM

A radiation worker receives a whole-body equivalent dose of 12 mSv, and he is also exposed to radioiodine, which results in an equivalent dose of 80 mSv to the thyroid. What additional whole-body equivalent dose could he still receive in order not to exceed the ICRP 103 recommendation dose limit?

Solution

$$\text{Effective dose} = \sum w_T H_T$$

where:

w_T = tissue weighting factor
H_T = equivalent dose to organ tissue T

According to ICRP 103 recommendation:

$$w_T \text{ for whole body exposure} = 1$$
$$w_T \text{ for thyroid} = 0.04$$

Hence, the effective dose received by the radiation worker is:

$$\sum w_T H_T = (1 \times 12 \text{ mSv}) + (0.04 \times 80 \text{ mSv}) = 15.2 \text{ mSv}$$

The ICRP 103 recommendation dose limit for radiation workers is 20 mSv per year.

$$20 \text{ mSv} - 15.2 \text{ mSv} = 4.8 \text{ mSv}$$

Thus, the worker could still receive a whole-body equivalent dose of 4.8 mSv in that particular year.

10.10 RADIOACTIVE WASTE MANAGEMENT

PROBLEM

Describe different methods of radioactive waste disposal.

Solution

i. Decay storage: This method is usually used for radionuclides with physical half-lives less than 120 days. The radionuclides are allowed to decay in storage for a minimum period of 10 half-lives. After 10 half-lives, if the radioactivity cannot be distinguished from the background activity, it can be disposed of in the normal waste after removing all the radiation labels.

ii. Release into sewage drain system: A small amount of water-soluble radioactive material may be disposed of in a designated sink, and the total amount of activity does not exceed the regulatory limits. Detailed records of all radioactive material disposals must be kept for inspection by regulatory bodies. Radioactive excreta from patients may be exempted from these limits and may be disposed into the sanitary sewer. Items that are contaminated with radioactive excreta (e.g. linen and diapers) have to follow the same limitations.

iii. Transfer of waste to an authorised recipient: This method is adopted for long-lived radionuclides and usually involves the transfer of radioactive waste to an authorised company for landfill or incineration at approved sites or facilities.

iv. Other disposal methods: The licensee may also adopt other disposal methods that are approved by the regulatory bodies. The impact of such methods on the environment, nearby facilities and the population must be justified before approval. Incineration of solid radioactive waste and carcasses of research animals containing radioactive material is allowed by this method. Radioactive gases (e.g. Xe-133) can be released by venting through a fume hood as long as their maximum permissible concentration does not exceed the regulatory limits.

10.11 DECONTAMINATION PRINCIPLES

PROBLEM

Describe the principles used in decontamination techniques.

Solution

i. Choose wet decontamination over dry decontamination, because the dry method could create airborne dust hazards.

ii. Use mild decontamination with non-abrasive agents first, as other methods could damage the surface involved.

iii. Take precautions to prevent further spread of contamination during decontamination operations.

iv. Isolate and separate contamination with short-lived activity to allow it to decay naturally.

10.12 OUT-PATIENT ADVICE

PROBLEM

What advice would you give to out-patients who have undergone low-dose I-131 treatment before leaving the hospital?

Solution

The following advice would be given to the patient:

- No drinking and eating during the first hour after treatment.
- Drink more water than usual during the following two days to stimulate urinary excretion.
- Use only one toilet if possible, and flush two to three times after each use.
- Wash hands frequently, and take a shower at least once a day.
- Avoid close contact with members of the family, especially children and pregnant women.
- Use disposable products (e.g. paper cups, disposable utensils and tissue paper) whenever possible.

10.13 RADIOIODINE WARD NURSING STAFF

PROBLEM

From the radiation protection perspective, what would be your advice to the nursing staff who are looking after the I-131 patients in radioiodine therapy wards?

Solution

Advice to the nursing staff would include:

- Spend minimum time with the patients by planning ahead and working efficiently.
- Work as far away as possible from the patients (inverse square law).
- Practice preventative measures against contamination, such as wearing disposable gloves and protective aprons.
- Remove personal protection clothing before leaving the room.

10.14 HANDLING OF THE I-131 PATIENT AFTER DEATH

PROBLEM

What are the precautions that are needed in an event of the death of a patient following I-131 treatment?

Solution

- According to the ICRP-94 recommendation, the activity limit for burial and cremation of patients following I-131 treatment is 400 MBq.
- The preparation for burial or cremation should be supervised by a competent authority or radiation protection officer (RPO).
- The funeral directors will need to be advised of any necessary precautions, and notification of the relevant competent authorities is required.
- Relatives and friends should not be allowed to come into close contact with the corpse.
- Clear communication is needed between the authorities, hospital staff, funeral directors and the family members to ensure that adequate controls are implemented without compromising dignity.
- All personnel involved in handling the corpse should be instructed by the RPO and monitored.
- All objects, clothes, documents, etc. that might have been in contact with the deceased must be monitored for any contamination.
- It may be expedient to wrap the corpse in waterproof material immediately after death to prevent the spread of contaminated body fluids.
- Embalming of the corpse should be avoided.
- Autopsy of a highly radioactive corpse should be avoided.

Bibliography

Bailey, D.L., Humm, J.L., Todd-Pokropek, A., van Aswegen, A., 2014. *Nuclear Medicine Physics: A Handbook for Teachers and Students*. Vienna, Austria: IAEA.

Cherry, S.R., Sorenson, J.A. and Phelps, M.E., 2012. *Physics in Nuclear Medicine e-Book*. Philadelphia, PA: Elsevier Health Sciences.

De Lima, J.J. Ed., 2016. *Nuclear Medicine Physics*. Boca Raton, FL: CRC Press.

Flower, M.A. Ed., 2012. *Webb's Physics of Medical Imaging*. Boca Raton, FL: CRC Press.

Griffiths, H.J., 1988. "Radiation dose to patients from radiopharmaceuticals." ICRP Publication 53. *Radiology*, 169(3), pp. 652–652.

International Atomic Energy Agency. *Nuclear Medicine Physics: A Handbook for Teachers and Students*. Vienna, Austria: IAEA, 2014.

International Commission on Radiological Protection, 1998. *Radiation Dose to Patients from Radiopharmaceuticals*. ICRP Publication 80, Addendum to ICRP 53. Oxford, UK: Pergamon Press.

International Commission on Radiological Protection, 2000. "Pregnancy and medical radiation." ICRP Publication 84. *Ann ICRP*, 30(1), pp. 1–43.

Lin, G.S., Hines, H.H., Grant, G., Taylor, K. and Ryals, C., 2006. "Automated quantification of myocardial ischemia and wall motion defects by use of cardiac SPECT polar mapping and 4-dimensional surface rendering." *J Nucl Med Technol* 34(1), pp. 3–17.

Powsner, R.A., Palmer, M.R. and Powsner, E.R., 2013. *Essentials of Nuclear Medicine Physics and Instrumentation*. Chichester, UK: John Wiley & Sons.

Russell, J.R., Stabin, M.G., Sparks, R.B. and Watson, E., 1997. "Radiation absorbed dose to the embryo/fetus from radiopharmaceuticals." *Health Phys.* 73(5), pp. 756–769.

Saha, G.B., 2012. *Physics and Radiobiology of Nuclear Medicine*. New York: Springer Science & Business Media.

Theobald, T. Ed., 2011. *Sampson's Textbook of Radiopharmacy*. London, UK: Pharmaceutical Press.

Printed in the United States
by Baker & Taylor Publisher Services